Mitigation of Hydrodynamic Resistance

Methods to Reduce Hydrodynamic Drag

Mitigation of Hydrodynamic Resistance

Methods to Reduce Hydrodynamic Drag

Marc Perlin • Steven Ceccio

University of Michigan, USA

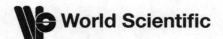

World Scientific

NEW JERSEY • LONDON • SINGAPORE • BEIJING • SHANGHAI • HONG KONG • TAIPEI • CHENNAI

Published by

World Scientific Publishing Co. Pte. Ltd.

5 Toh Tuck Link, Singapore 596224

USA office: 27 Warren Street, Suite 401-402, Hackensack, NJ 07601

UK office: 57 Shelton Street, Covent Garden, London WC2H 9HE

Library of Congress Cataloging-in-Publication Data
Perlin, Marc.
   Mitigation of hydrodynamic resistance : methods to reduce hydrodynamic drag / by Marc Perlin (University of Michigan, USA), Steven Ceccio (University of Michigan, USA).
     pages cm
   Includes bibliographical references and index.
   ISBN 978-9814612258 (alk. paper)
   1. Ships--Hydrodynamics. 2. Ship resistance. 3. Hulls (Naval architecture) I. Ceccio, S. L. (Steven L.) II. Title.
   VM751.P44 2015
   623.8'1015325--dc23

                                      2014031056

British Library Cataloguing-in-Publication Data
A catalogue record for this book is available from the British Library.

In-house Editors: Sutha Surenddar/Steven Patt

Typeset by Stallion Press
Email: enquiries@stallionpress.com

Printed in Singapore

To Terry and Martha

Preface

This text is written for scientists and engineers who are interested in various aspects of, and techniques to reduce the drag forces on objects moving at or below a liquid surface. Although there is a combination of experimental, numerical, and analytical material presented, the dominant focus is experimental in nature. Numerical investigations have aided understanding of the underlying physical mechanisms associated with flow induced skin friction and its reduction; however, numerical efforts are often limited by the computational resources needed to simulate large Reynolds number flows at large scales. That is, a typical surface ship in the ocean environment operates at Reynolds numbers of about 10^9 with lengths of 100 m or more, hence the focus on experiments. As always, the best approach is a thoughtful combination of experiment, computation, and analysis.

Senior undergraduate students who have successfully completed a course or more in fluid mechanics and graduate students, who have done likewise, can master the material in this book. Practicing engineers who desire knowledge in this area will also find the material acceptable and useful as well as easily grasped and mastered.

Acknowledgments

To all of the Graduate students who have facilitated this research, we acknowledge your dedication and extremely difficult efforts at the US Navy's Large Cavitation Channel and at the University of Michigan. Experimental research is very difficult to conduct, especially at these large scales and expensive facilities. These students include Dr. Wendy Sanders, Dr. Eric Winkel, Dr. Brian Elbing, Dr. Ghanem Oweis, Dr. Simo Makiharju, Dr. Keary Lay, Dr. Andrew Wiggins, Dr. Xiaoshun Shen, Dr. Jinhyun Cho and Dr. Ryo Yakushiji, and Mr. James Gose and Ms. Sarah Schinasi.

To our colleagues, Professors David R. Dowling and Michael J. Solomon, we thank you for your collaboration, effort, and technical insights.

Thanks to our colleagues at the Naval Surface Warfare Center, Carderock Detachment, especially Mr. Robert Etter, and the staff at the Memphis Detachment, especially Dr. Michael Cutbirth, H. Paul Julian, and all the many technicians at the Large Cavitation Channel. As did our colleagues at UM, you made our work successful there and as much fun as it could be under the strenuous conditions and long hours.

We are indebted to the sponsors of many of these investigations, the Defense Advanced Research Project Agency, and the Office of Naval Research. We would like to thank Dr. Lisa Porter, Dr. Thomas Buetner, Dr. Pat Purtell, and Dr. Ki-Han Kim.

Lastly, we thank the University of Michigan for the opportunity to prosper and conduct our research as we choose. To date, both of us have spent our entire academic careers there, and for this we are appreciative.

Contents

Chapter 1

Introduction

All types of vehicles encounter resistance to their (forward) motion, and minimizing this resistance has been a goal of engineers and designers for time immemorial. Hydrodynamic resistance, for example, of surface and sub-surface vehicles can broadly be separated into the following physical processes: (1) skin friction drag, (2) form drag, (3) wave-making resistance, and (4) induced drag. Designers strive to minimize the total resistance of a vessel while maintaining desired operational characteristics. However, the minimization of one form of resistance often leads to an increase in another. For example, the use of long and slender hulls can reduce both form and wave drag but may lead to an increase in skin friction drag.

Under ideal circumstances, a vessel's resistance can be reduced *via* passive means. That is, the shape and surface of the hull is designed to reduce the drag force without additional input of mass, momentum, or energy beyond that which is normally required to propel the ship. However, it may be both practical and advantageous to modify the resistance through the *active* input of mass, momentum, or energy in the flow near the ship hull. For these methods to succeed, the *costs* associated with active flow modification must be outweighed by the *benefit* in reduced propulsion power or increased performance.

Successful application of passive drag reduction (DR) includes the installation of bulbous bows. The bulbous bow has a long history beginning with D.W. Taylor around the turn of the 20th century, and continued/improved by Havelock (1934), Saunders (1957), Wigley (1936), and in the early 1960's made popular by Inui, according to the text *Principles of*

Naval Architecture, Volume II. The presence of the bulb at the bow of the ship produces bow waves that cancel ship-generated waves that reduce the overall wave drag. And, the energy savings of a properly designed bulb far outweigh the additional friction drag on the bulb surface.

Another basic method of passive skin friction reduction is the application of paints and other coatings on the hull. Surface coatings can be applied to a hull to decrease the level of surface roughness and to inhibit the formation of biofouling, thus decreasing the skin friction drag. Any effort to smoothen the hull directly impacts the ship resistance. Schultz (2007) reported that the powering requirements of a cruising navy frigate freshly coated with anti-fouling paint is expected to be only 4% higher that the drag of the same ship with an ideally (i.e. hydraulically) smooth surface. However, the presence of slime increases the powering required by 10% to 16%, while the presence of calcareous fouling increases the powering by 26 to 59%. Therefore hull maintenance as regards smoothness is the first line in passive skin friction reduction.

Chambers *et al.* (2006) review different types of anti-fouling coatings, including coatings impregnated with biocides (typically heavy metals). Self-polishing paints that contained a copolymer matrix of antifoulant tributyltin (TBT) and copper are very effective; however the effect of TBT on the environment has led to its prohibition. As an alternative, newer ablative coatings are being developed. These "foul release coatings" have a low surface energy that promotes the detachment of bio-fouling under flow induced shear. The development and application of anti-fouling coatings is of extreme importance to resistance reduction. And, while we do not intend to review the field in this chapter, the reader is referred to Yebra *et al.* (2004) and Chambers *et al.* (2006) for an introduction to this topic. Highly engineered surfaces with structured roughness can also be applied to the hull to reduce friction. The best known example is "riblet surfaces," which are small-scale, longitudinal grooves used to produce local shear reduction (see e.g. M.J. Walsh papers and Bushnell and McGinley, *Annu. Rev. Fluid Mech.*, 1989). We will review some of the more well-known methods of passive DR in this text.

Active techniques of DR require continuous engagement to achieve results. Many such methods exist including the addition of polymers or fibers into the boundary layer of the flow around the hull, gas/bubble injection, the formation of air or vapor cavities around a lifting surface and air layers beneath hulls. These methods are primarily used to reduce skin friction. Flows over solid surfaces develop drag as a result of the no-slip condition, making friction drag an important consideration for aerodynamic and hydrodynamic systems. Friction drag is of particular importance for marine system, where roughly 50% of a ship's and 60% of a submarine's drag results from skin friction. And operation in the marine environment can lead to a tremendous increase in friction drag as the hull roughness increases as a result of fouling. Therefore, reduction in skin friction can lead to immediate decreases in fuel consumption (and operating costs) along with increases in performance. There has been recent renewed interest in the use of gas injection to reduce friction drag around surface ships, with successful applications being reported by groups in Europe and Asia. And, these methods are now being commercialized. Active methods are the principal topic of this manuscript.

Development and adoption of novel resistance reduction methods is made challenging by the standard design process of large-scale marine vehicles. Once the hullform and propulsion systems have been designed, it is common practice to test the hydrodynamic performance of the vehicle at model scale. Then, the model scale results are used to infer full-scale performance. Froude and Reynolds number scaling of the passive hydrodynamic resistance from model tests is well established for conventional hullforms. However, not all resistance reduction methods can be easily scaled, especially those that rely on modification of high Reynolds number boundary layers. In fact, we will discuss how scale effects can be an important consideration for the application of active DR techniques.

It is therefore beneficial to first understand the basic physical processes that lead to reduction in drag. In doing so, it is then possible to develop analytical and numerical methods that can aid in the scale-up of candidate methods. And, it is important to collect test data at the highest Reynolds number possible on generic test models. Doing so reduces the risk associated with the application of novel DR methods to full-scale marine vehicles. Much of the material presented here is based on research guided under these principles.

Herein we explore our (i.e. the authors' and their collaborators, of whom Professors Dowling and Solomon are primary) own explanations and view of DR. The reader should note that there are myriad references on the general subject of DR, although the ones used in this text are those primarily from large-scale experiments conducted by the authors.

As far as the definition of DR, we usually report and discuss a %DR defined as

$$\%DR = \left(\frac{\tau_{w/o} - \tau_{w/}}{\tau_{w/o}} \right) * 100 = \left(1 - \frac{\tau_{w/}}{\tau_{w/o}} \right) * 100$$

where $\tau_{w/}$ is the shear stress with active or passive modification and $\tau_{w/o}$ is the shear stress without modification.

To close our introductory remarks, we comment that the list of references given in this text should be invaluable to the novice. Obviously they make for important reading.

References

Bushnell, D.M. and McGinley, C.B. (1989). Turbulence control in wall flows. *Annu. Rev. Fluid Mech.* 21, 1–20.

Chambers, L.D., Stokes, K.R., Walsh, F.C. and Wood, R.J.K. (2006). Modern approaches to marine antifouling coatings. *Surface Coatings Tech.* 201, 3642–3652.

Havelock, Th.H. (1934). Wave patterns and wave resistance. *TINA*, 76, 430–446.

Saunders, H.E. (1957). Hydrodynamics in ship design. *SNAME*, I.

Schultz, M.P. (2007). Effects of coating roughness and biofouling on ship resistance and powering. *Biofouling*, 23(5), 331–341.

Wigley, W.C.S. (1936). The Theory of the Bulbous Bow and its Practical Application. North East Coast Institution of Engineers and Shipbuilders, 52.

Yebra, D.M., Kill, S. and Dam-Johansen, K. (2004). Antifouling technology — past, present and future steps towards efficient and environmentally friendly antifouling coatings. *Prog. Organic Coatings* 50, 75–104.

Part I

Active Techniques

Chapter 2

Modification of Drag by Polymer Injection

2.1. Background Information

For many persons using this text, background on polymer mixtures and inner boundary layer processes may be lacking; hence an effort is made to cover information that we feel may be required. We begin with a discussion of wall-bounded, turbulent shear flows. This material follows closely the text *Fluid Mechanics* by Kundu *et al.* (2012), and the interested reader is referred there for additional information not covered here.

2.1.1. Inner variables and scaling

Wall-bounded turbulent shear flows include the so-called law of the wall, inner variables, the viscous sublayer, the outer profile and the overlap region for example. At sufficiently large Reynolds number (Re) free turbulent shear flows become Re independent; however in the vicinity of a solid surface/wall, viscosity is always important. In the immediate vicinity of a wall, another inner length scale arises, l_ν, the viscous wall unit or viscous length scale associated with the viscous sublayer or inner layer. This scale complements the usual outer length scale, the boundary layer thickness, δ. In these flows the friction effects remain a function of Re. In Figure 1 taken from Kundu, Cohen, and Dowling (their Figure 12.18) via data from Oweis *et al.* (2010), mean velocity profiles are shown over a smooth flat plate for downstream-distance-based Reynolds numbers, Re_x, of 0.73 to 2.20×10^8. These are equivalent to momentum-thickness-based Reynolds numbers, Re_θ, of 56,000 to 150,000 for their experiments. On the graph, the abscissa is the distance from the wall, y, non-dimensionalized by the viscous length scale and is known simply as y^+, while the ordinate scale is known as the shear or friction velocity, u_*, herein, and is the velocity, u, which has been non-dimensionalized by the square root of the quantity — the shear at the wall, τ_w, divided by the mass density, ρ.

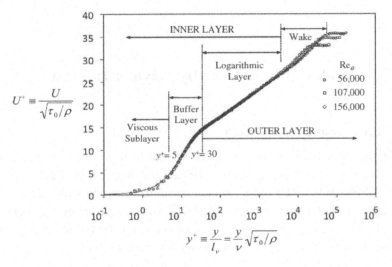

Figure 1. Mean velocity profiles for turbulent flow over a smooth flat plate at large Re.

Subdivision of the turbulent boundary layer (TBL) is based on scaling: (1) the "law of the wall" governs the region where viscosity is important and this region is called the "inner layer" as can be seen in the figure, here l_ν is the most important length scale; (2) the second scaling law is the "velocity defect law". This region of the flow is known as the "outer layer" and it is where the flow is largely independent of ν, and δ is the relevant length scale. Note that all of the data collapse in the scaled space of Figure 1.

Let us assume that u represents the mean velocity ($u = \bar{u} + u'$ where as usual the over bar represents the mean and the prime represents the fluctuating part, and we drop the over bar for simplicity) in the flow direction at any location; then, near the wall, only the kinematic viscosity, ν, the mass density, ρ, the distance from the wall, y, and the shear stress on the wall, τ_w, should affect u. Equivalently $u = u(\rho, \nu, y, \tau_w)$. Dimensional analysis then gives

$$u^+ \equiv \frac{u}{u_*} = f\left(\frac{u_* y}{\nu}\right) = f\left(\frac{y}{\nu/u_*}\right) = f\left(\frac{y}{l_\nu}\right) = f(y^+)$$

where $u_* = \sqrt{\tau_w/\rho}$, the so-called friction or shear velocity, and $l_\upsilon = \frac{\upsilon}{u_*}$ is the viscous wall unit. Very close to the wall, the upper limit of the viscous sublayer is $y^+ \approx 5$ (see Figure 1). And using the 2D Newtonian fluid assumption, $\tau_w = \mu \frac{\partial u}{\partial y}$, and assuming that the sublayer is linear means $\frac{\partial u}{\partial y} \approx$ constant for large Re. This implies then that $u^+ = y^+$.

For the outer layer, we have stated already that the flow is essentially independent of υ; hence it is "inviscid-like". Thus in the outer layer, $u = u(\rho, \delta, y, \tau_w)$, but u is still less than u_∞ (i.e. u of the free stream). The velocity defect, $u_\infty - u$, then by dimensional analysis yields

$$\frac{u_\infty - u}{u_*} = g\left(\frac{y}{\delta}\right) \equiv g(\eta)$$

or $u_\infty - u = u_* g(\eta)$ with η the usual boundary layer coordinate and this last expression known as the velocity defect law (outer layer).

As can be seen in Figure 1, the inner and outer layers overlap, and thus this region of intersection is known as the overlap region. In the inner region, y scales by l_υ, in the outer region by δ. If we allow $y^+ \to \infty$ and $\eta \to 0$ (i.e. allow asymptotic matching), we obtain the overlap region.

It turns out that the simplest matching is to match slopes (velocity gradients), du/dy. Using the law of the wall $\left(\frac{u}{u_*} = f\left(\frac{u_* y}{\upsilon}\right)\right)$ and the velocity defect law $[u_\infty - u = u_* g(\eta)]$, taking the derivative, $d(u)/dy$, yields the inner and outer derivatives, respectively,

$$\frac{du}{dy} = u_* \frac{df}{d\left(\frac{u_* y}{\upsilon}\right)} \frac{d\left(\frac{u_* y}{\upsilon}\right)}{dy} = \frac{u_*^2}{\upsilon} \frac{df}{dy^+}$$

$$\frac{du}{dy} = -u_* \frac{dg}{d\eta} \frac{dn}{dy} = -\frac{u_*}{\delta} \frac{dg}{d\eta}$$

and equating gives

$$y^+ \frac{df}{dy^+} = -\eta \frac{dg}{d\eta}.$$

Consequently as the left hand side of the equation is a function only of y^+ and the right hand side is a function only of η, each side must equal the same separation constant which taken with hindsight is $1/\kappa$. Empirically, it has been shown that $\kappa \approx 0.4$ and it is known as the von Karman constant.

Separating the variables now and integrating gives the velocity profile in the overlap region or the so-called logarithmic layer:

$$f(y^+) = \frac{1}{\kappa} \ln(y^+) + B \quad \text{and} \quad g(\eta) = -\frac{1}{\kappa} \ln(\eta) + A.$$

Empirically, it is found that $4 < B < 5$ and $A \approx 1$ for an ideally smooth wall. Note that as these equations are valid only in the overlap region, they are for large y^+ and small η.

From a previous equation we had

$$u^+ = \frac{u}{u_*} = f(y^+)$$

which allows the first of the two equations above to be written as

$$\frac{u}{u_*} = u^+ = f(y^+) = \frac{1}{\kappa} \ln(y^+) + B$$

known as the logarithmic- or log-law. Much of the following information on polymer drag reduction (PDR) uses inner coordinate scaling.

2.1.2. Background for PDR

Returning to drag reduction (DR) using polymer injection, we require yet additional background, specifically regarding friction factors such as those due to Darcy and to Fanning, and knowledge of Virk's maximum drag reduction (MDR) asymptote. Likewise, we mention the manner in which we display the data, i.e. using the Fanning friction factor (recall that $f = 16/\mathrm{Re}$, and is equal to $1/4^{\text{th}}$ of the Darcy friction factor, $f_{\text{Darcy}} = 64/\mathrm{Re}$). As the Fanning friction factor uses ρu_∞^2 or ½ this value to non-dimensionalize the stress, and higher speed drag generally is non-dimensionalized by $\frac{1}{2}\rho u_\infty^2 A$, it is most often used as it will be herein with $\frac{1}{2}\rho u_\infty^2$.

Next, we discuss the onset of PDR. This discussion follows loosely that of White and Mungal (2008). Following Toms (1948) publication, experimental evidence demonstrated that laminar flow in pipes with low polymer concentrations did not exhibit DR, but turbulent flow friction could be significantly reduced with the addition of high molecular-weight polymers in solution (As is well-known, turbulence dynamics are a function of Re, while polymer dynamics are a function of the number of monomers in the macromolecule, i.e. in the polymer. A monomer is defined as a low molecular weight (MW) molecule capable of reacting with identical or different molecules also of low MW to form a polymer that has high MW. Thus one can think of monomers as the building blocks of polymers. And, as the experiments showed that for a given pipe diameter and flow, the DR required a particular number of monomers, there is a threshold concentration required to obtain PDR.).

There are generally two accepted explanations for PDR onset, both based on the effects on the flow of polymer stretching. One group argues that viscous effects are germane while the other camp's argument is based on elasticity. The first group argues that the stretching of the polymer essentially causes an increased extensional viscosity (this is somewhat analogous to the idea that the turbulence caused by surface water wave breaking effectively increases viscosity and thereby increases wave attenuation with propagation through regions of turbulence caused by breaking), and that this increased viscosity suppresses turbulent fluctuations and thus reduces drag (This general notion, that the reduction of turbulent fluctuations due to whatever mechanism is responsible for DR, is and will be a recurring explanation for DR whether polymer or not.). Formal models have been derived by several researchers.

The other group of researchers base the PDR on an elastic theory where the elastic energy stored in the polymers is order of the kinetic energy in the buffer layer, and that the energy cascade is disrupted, thus causing a thicker buffer layer and hence DR. Experiments show that each theory is sensible, but as elasticity and viscosity are somewhat coupled, the actual answer is likely a combination of both processes.

The next idea that is required is that of an upper bound for PDR, the so-called Virk asymptote that we will graph usually in Prandtl-Kármán form. For pipe and other frictional flows, as is well-known, friction drag is a function of the Reynolds number. Recall that Re is a dimensionless number formed as a ratio of the inertia to viscous forces, or that $Re = UL/\upsilon$ with U, L, and υ a characteristic speed, length, and the kinematic viscosity, respectively. For example, a Re number for pipe flows might have U the mean flow speed and the characteristic length, L, as the pipe diameter. Other choices exist such as the maximum flow speed (i.e. along the pipe centerline) and the pipe radius for the length scale. Using the Fanning friction factor for fully-developed pipe flow, we have $f = \frac{\Delta p}{\Delta x} \frac{D}{2\rho U^2}$ with Δp the pressure drop along the pipe and Δx the pipe length. We can graph the relationship between Re and f with Prandtl–Kármán (PK) coordinates, employing $\text{Re}\sqrt{f}$ on the abscissa and ordinate $1/\sqrt{f}$. Superposed on these coordinate then is the pipe friction relationship for a Newtonian fluid, the so-called PK law:

$$\frac{1}{\sqrt{f}} = 4.0 \log_{10}\left(\text{Re}\sqrt{f}\right) - 0.4.$$

The MDR asymptote or Virk asymptote (Virk *et al.*, 1967), provides an upper limit on PDR and is given by Virk and co-workers as

$$\frac{1}{\sqrt{f}} = 19.0 \log_{10}\left(\text{Re}\sqrt{f}\right) - 32.4.$$

If the polymer concentration is increased beyond that value required to achieve this limit, the shear viscosity increases and hence friction then begins to increase (rather than to decrease further). Also important, and presented in Figure 2 using curve fits, is the polymeric flow regime as given by Virk:

$$\frac{1}{\sqrt{f}} = (4.0 + \delta) \log_{10}\left(\text{Re}\sqrt{f}\right) - 0.4 - \frac{\delta}{L} \log_{10}\left(\sqrt{2}DW^*\right),$$

where δ is the so-called slope increment, and W^* is the so-called onset wave number $\left(\equiv \frac{1}{v}\sqrt{\frac{\tau_w^*}{\rho}}\right)$; both are polymer solution dependent parameters (i.e. a function of the polymer and its solvent along with its MW and concentration) and τ_w^* is the wall onset shear stress. Figure 2 represents an experimental result and is taken from Garwood *et al.* (2005) (their Figure 7).

At this point, the figure is presented to aid a discussion of the PK graph and curve along with the MDR, and the other Virk equations just presented to aid the reader with becoming familiar with the presentation manner. Hence, generally increasing the ordinate indicates that friction is decreasing or that there is more DR. However note that as the abscissa, $\text{Re}\sqrt{f}$, also includes the friction factor, increasing the ordinate due to a decrease in f causes the abscissa to decrease simultaneously. For fixed Re, higher polymer concentration results in lower f (increased DR), but note that as you jump from one concentration to another at constant Re, since f decreases, you move up and to the left in the figure. In addition, for a given concentration curve, when the curve goes from straight to curved in the figure, chain scission to be discussed later is the likely cause. A last comment is that the data-fitted lines on the figure are fitted only through the straight portion of those data (see Garwood *et al.* (2005) for a discussion).

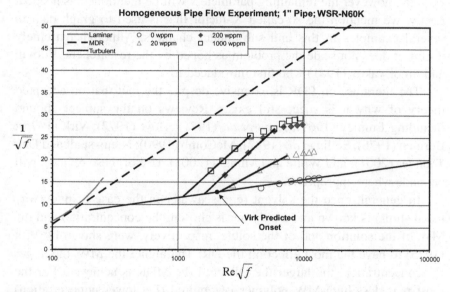

Figure 2. Figure from Garwood *et al.* (2005) graphed on PK coordinates showing the MDR, the PK curve (denoted Turbulent in the figure key), the laminar results for water, and four concentrations of the polymer (WSR-N60K, a Dow polyethylene-oxide [PEO] polymer [WSR is an acronym that indicates Water Soluble Resin] that has a MW of 2.3×10^6 g/mole, usually taken as ≈ 2M) — water mixture. The designation wppm indicates weight-parts-per-million.

From a practical point of view, as noted previously, the Trans-Alaskan pipeline has used polymer to reduce pumping costs and/or to reduce pumping stations since its inception. And, as mentioned also, the U.S. Navy has not used polymers for DR on its ships due to the cost of the polymer, the capital and maintenance costs of the infrastructure to inject, and the reduced payload/cargo of the vessel due to the necessity of storing the material. The reader should note further that the MW and the concentration

(c) of the polymer mixture play an important role in DR technology and that scission/degradation of the polymer by turbulence, pumping, etc. cause a decrease in MW, thus reducing DR.

Before we delve into methodology and results, some other parameters and dimensionless numbers are mentioned: *We*, Weissenberg number; $[\eta]$, intrinsic viscosity; Zimm or relaxation time; Θ_Z, a relationship between DR onset's minimum shear rate, γ^*, and the critical rate for shear degradation, γ_D; etc.; however the remaining parameters will be mentioned as needed. Lastly, we mention the so-called K-factor that is used to graph data in several examples in this and subsequent chapters. Although extremely useful, it does not scale the problem as hoped by the researchers, Vdovin and Smol'yakov (1981) who first introduced it.

The literature on PDR is extensive despite the fact that no accepted theory of why it is successful exists. Reviews on the subject abound including Lumley (1969); Liaw *et al.* (1971); Hoyt (1972); Virk (1975); Berman (1978); Sellin *et al.* (1982); McComb (1990); Nieuwstadt and Den Toonder (2001); and White and Mungal (2008). Use of these reviews will be made where appropriate.

In general, once the solvent (e.g. seawater in the case of underway naval ships) is known and a polymer is chosen, the concentration and the MW of the solution and of the solute, respectively, were shown by Virk (1975) to have the most effect on the DR. The higher the MW, the larger the concentration, the larger the DR until the MDR is achieved. For the most part, less high-MW polymer is required (i.e. lower concentration) to achieve the same DR as compared to that required with a lower MW material. Hence it may be cost-effective to use high-MW polymer but at a lower concentration. The difficulty is that higher MW polymer is made of longer chain molecules, thus the higher the likelihood of so-called chain scission (see e.g. Patterson and Abernathy, 1970), that is degradation by breaking or dividing the chain. Until recently, the investigation of scission focused primarily on laminar flows. The study by Vanapalli *et al.* (2006) demonstrated that the degradation of the polymer occurs at the Kolmogorov turbulence length scales. In that study a universal scaling was proposed; here without derivation we will state and use their results where required. The interested reader is referred to their paper.

2.2. Internal Flows at High Re

During the last decade or so, the authors of this text along with several of their collaborators have undertaken investigations of internal flows that were necessary to determine information required to investigate external flows that were of direct interest. In a forerunner to Elbing *et al.* (2009), a paper that will be discussed at length, Garwood *et al.* (2005) investigated the flow of polyethylene oxide or polyox (PEO) at high shear rates (to $3 \times 10^5 \text{ s}^{-1}$) and Re to 4×10^5. MW in these experiments ranged from 2×10^6–4×10^6 g/mole (and even 8×10^6 in earlier tests) while concentration was in the range of 10–1000 wppm. The experimental setup in both investigations was essentially the same with more pressure taps in the earlier study but additional sampling ports added in the later investigation. The setup for Elbing *et al.* (2009) is shown in Figure 3 (Figure 2 of their paper).

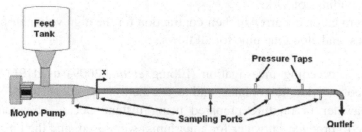

Figure 3. The experimental setup of Elbing *et al.* (2009). The experimental setup of Garwood *et al.* (2005) was similar with additional pressure taps, but no sampling ports.

A gradual transition section as shown on the upstream end had a 10.5° included angle that reduced the diameter from 20.3 cm (8 in) to the test section diameter of 2.7 cm (i.e. so-called 1 in pipe), but even at this relatively slow convergence, the two sampling ports spanning the transition exhibited degradation. In the earlier paper, a 5.24 cm (2 in pipe) was used for the long section as well as the 1 in pipe already mentioned. The 2.7 cm inner diameter produced flows with shear rates as in the Large Cavitation Channel (LCC) experiments that will be discussed subsequently while the 5.24 cm pipe had boundary layer thicknesses similar to those same experiments. For the Garwood *et al.* tests, and for most of our experiments in the LCC, three PEO polymers were used that included nominal MWs of 2M, 4M, and 8M. Concentrations pumped were 10, 20, 100, 200, and 1000 wppm while the mean flow speeds were 16 and $7 \, \text{ms}^{-1}$ in the smaller and larger pipes, respectively.

It should be reiterated here that strict mixing procedures and stability of the solutions are extremely important. Mixing should be on the order of one day with a mixed shelf-life of order three days, but testing and comparison to MDR insures that the polymer solutions are effective. Ethanol is used often in the dissolving process, although it is not absolutely necessary, but chlorine that degrades PEO must be removed for example by filtering through activated carbon prior to mixing the polymer with water containing chlorine, as in city water. The primary conclusions of Garwood *et al.* (2005) are:

(1) Onset of DR is a function of concentration; this in fact disagrees with the findings of Virk.
(2) Degradation occurred in their contraction for the high volumetric flow rates, and along the pipe for all flows.

The succeeding investigation (Elbing *et al.*, 2009) used PEO and PAM (polyacrylamide) solutes, and tested them in turbulent flows. Water and seawater (a mixture prepared using Instant Ocean, an aquarium additive to produce saltwater for aquariums) were used, and the universal scaling law of polymer scission (Vanapalli *et al.*, 2006) was extended for greater Re. Reiterating their universal scaling law mentioned earlier, they showed definitively that scission occurs on Kolmogorov length scales (the smallest turbulent or eddy length scales, $\eta = \left(\frac{\nu^3}{\bar{\varepsilon}}\right)^{1/4}$, where ν is as usual the kinematic viscosity and $\bar{\varepsilon}$ is the average rate of kinetic energy dissipation per unit mass supplied to the smallest eddies). Additionally as DR implementation on ships traveling in seawater is an important application, this study repeated the filtered tap water experiments with the pseudo-seawater as regards DR and degradation. Some of their most important findings are included with their Figures 5, 6, and 8 presented as Figure 4.

The first of these figures (the three graphs of their Figure 5) shows the PK plots for all concentrations tested for the three PEO solutions tested. As usual the Newtonian friction line and the MDR are shown also. All data

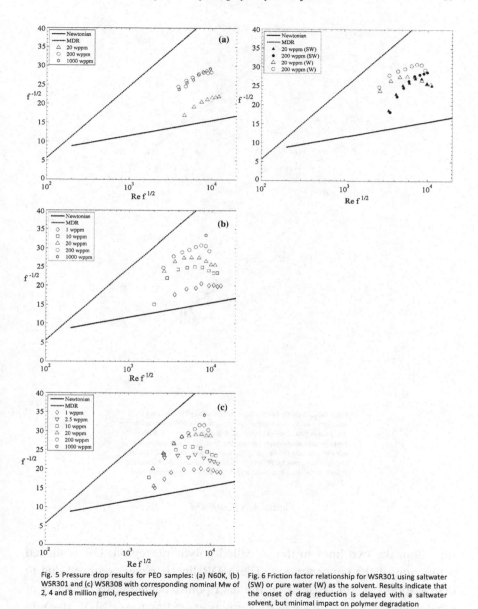

Fig. 5 Pressure drop results for PEO samples: (a) N60K, (b) WSR301 and (c) WSR308 with corresponding nominal Mw of 2, 4 and 8 million gmol, respectively

Fig. 6 Friction factor relationship for WSR301 using saltwater (SW) or pure water (W) as the solvent. Results indicate that the onset of drag reduction is delayed with a saltwater solvent, but minimal impact on polymer degradation

Figure 4. Figures 5, 6, and 8 from Elbing *et al.* (2009) that show pressure drop results for the three PEO MWs, friction factor results for the 4M MW polymer solution with fresh and with salt water, and F_{max} dependence as a function of Re.

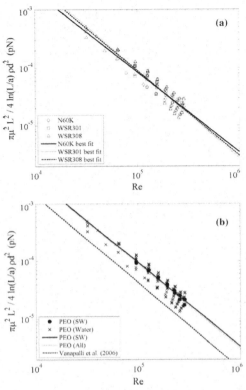

Fig. 8 F_{max} dependence on Re; a plotted with all results for PEO solutions having a pure water (W) solvent and b PEO solutions with a saltwater (SW) solvent plotted with pure water results. Both plots include least-squares power-law best-fit curves, which were used to estimate the Reynolds number dependence. Lines in b are the best-fit curves for PEO with saltwater solvent, all PEO results from the current experiment and PEO results from Vanapalli *et al.* (2006)

Figure 4. *(Continued)*

lie within the two lines in the so-called polymeric region. DR is indeed a function of the concentration. The MDR line is never achieved due to degradation (as can be seen by the data points deviating to the right of what would be a straight set of data points extending to the MDR line). In their Figure 6, two concentrations are given for a filtered water solvent and an Instant Ocean solvent. It is clear from constructing a line using the data points for each concentration–solvent mix that the onset of DR increases for saltwater (i.e. that it is delayed to higher Re). However, the slope increment and the degradation are somewhat less affected (For the same concentration, the onset point moves to the right and down with saltwater, i.e. toward the Newtonian friction line, whereas one wants the point to move to the left and up in the figure as much as possible toward the MDR line.).

The interpretation and description of their Figure 8 requires additional background. Using the results of Vanapalli *et al.* (2006), Elbing *et al.* (2009), who were unable to relate MW (i.e. changes due to degradation) directly to Re, chose another approach — they applied the universal scaling law for polymer chain scission found by the former authors to their higher Re (and thus higher shear rate) data. This assessed the scaling law at much different scales.

As mentioned Vanapalli *et al.* argued that the degradation occurs at the Kolmogorov scales and determined that the maximum force on a polymer chain was:

$$F_{max} = \frac{A^{3/2}\pi\mu^2\text{Re}^{3/2}L^2}{4\rho D^2 \ln(L/a)}$$

where A is a flow geometry-dependent constant of proportionality, μ is the solvent dynamic viscosity, L is the extended length of the polymer, and "a" is the diameter of the polymer. Boyer and Miller (1977) published results that a for PEO is approximately 10^{-9} m. The value of L is problematic for several reasons including that the polymers were "polydisperse" (as opposed to monodisperse where all have the same MW), their usual coiled state may not be completely uncoiled and thus elongated, etc. However proceeding on the assumption that the L-values could be assumed to be order-of-magnitude correct, Vanapalli *et al.* estimated L via a relation published in Larson (1999), $L = 0.82\,nl_o$. Here $n = n_o M_{WS}/M_o$, the number of so-called backbone bonds, n_o is the number of backbone bonds per monomer, M_{WS} is the steady-state molar mass for scission in a wall-bounded fixed shear-rate flow, and M_o is the monomer molar mass. In addition l_o is the mean backbone bond length (Recall that a monomer is the molecule from which the polymer is prepared, while backbone bonds are the material and bond that holds the polymer together. For additional reading on backbones, we suggest the website http://ghr.nlm.nih.gov/handbook/basics/dna that discusses DNA and the notion of backbones.). PEO has a monomer backbone of C–C and two C–O bonds (with lengths l_o of 1.54 Å and 1.43 Å, respectively) for a total of three (i.e. $n_o = 3$), and its molar mass is 44.1 g/mol.

To complete the analysis, additional assumptions were required. First, they assumed that all PEO bonds break at the same F_{max}; then Re and L are related. To obtain this relationship, they began with the wall shear rate that is proportional to τ_w, and of course τ_w and f are related $\left(f = \frac{\tau_w}{\frac{1}{2}\rho U^2} = \frac{\Delta p}{\Delta x} \frac{D}{2\rho U^2} \right)$. Using that the shear rate is τ_w/μ, its value was found from the data. Additionally, they required a result from Vanapalli *et al.* (2005) where the wall shear rate was found to be $3.4 \times 10^{18} M_{WS}^{-2.20}$. From these they generated the last inset of our Figure 4 (Elbing *et al.* (2009), Figure 8) with the ordinate equivalent to $F_{max}/\left[A^{3/2} Re^{3/2} \right]$ as a function of Re. Both treated water and pseudo-seawater are presented.

Using these results by determining α in $F_{max} \propto Re^{\alpha}$, the researchers found $\alpha = 1.39, 1.52, 1.47,$ and 1.45 for the three polymers in water and for WSR-301 (4M MW) in salt water, respectively. Averaging these results gives $\alpha = 1.46$ which is very close to 1.5 as per Vanalpalli *et al.* (2006). Furthermore, A, the proportionality constant found by Vanapalli *et al.* was 2.09 ± 1.15, and so to obtain an order of magnitude approximation of breaking strength, $A = 1$ was used. This yields 3.2 ± 1.2 nN that is close to the published values of C–C and C–O bonds of 4.1 and 4.3 nN, respectively. The authors addressed the PAM results similarly. Thus the universal scaling law for scission at larger Re was supported.

Having now established some background ideas for polymers regarding scission/degradation, onset of DR, and the MDR used in the PK plot for internal flows, we can proceed to our primary flows of interest — external boundary layer flows with PDR.

2.3. External Flows and PDR

The authors and collaborators first published results on PDR in external turbulent flows appeared in FAST 2005, and were improved subsequently and appeared in the *Journal of Fluid Mechanics* in 2009, 621. This comprehensive article is discussed thoroughly herein.

In these experiments conducted at the U.S. Navy's William B. Morgan LCC, speeds to 20 ms^{-1} were used to measure friction drag reduction (FDR) via injection of PEO into TBLs. Six floating plate force balances were used to quantify the DR on a flat plate subjected to flows with Re numbers to 220×10^6 and varied concentrations and injection rates. Planar laser-induced fluorescence (PLIF) systems were used to determine concentrations in the near-surface, downstream flows. Prior to this set of experiments, studies on flat plates in the presence of turbulent flows used PEO and PAM and achieved Re to 45M (Vdovin and Smol'yakov, 1981) and speeds to 16.8 ms^{-1} (Petrie *et al.*, 1996). Also, FDR reduction of as much as 70% had been demonstrated.

For a brief introduction to the latest thinking on how polymers likely modify the flow to reduce drag, the reader is referred to Dubief *et al.* (2004). To arrive at the conclusions presented in that publication, the so-called Finite-Elastic Non-linear Extensible — Peterlin (FENE-P) model was used to simulate various flows with polymers. They discuss in detail how the elongation and recoiling of polymer molecules adjacent to a wall can modify the usual near-wall turbulence cycle as regards the buffer-layer vortices and sub-layer streaks. Their explanation is that in the near-wall region the polymer extracts energy from the vortices and releases it to the streaks leading to increased stream-wise velocity fluctuations, reduced wall-normal velocity fluctuations, and larger stream-wise vortex spacing. A combination of viscous, Re, and polymer stresses are thought to provide the near-wall shear stress balance. Additionally it is shown that MDR is achieved when the near-wall turbulence regeneration cycle is again self-sustaining due to the availability of sufficient polymer. As stated earlier herein, at this point additional polymer does not produce further DR, but rather causes an increase in shear viscosity.

2.3.1. The experiments of Winkel *et al.* (2009)

Returning to a discussion of the LCC experiments, to understand scale effects so as to facilitate scaling to ocean-going vessels with Re to $O(10^9)$, one requires knowledge of the mixing of the polymers, its degradation, and how it scales (in this case to experiments at smaller scales). In particular, you need to know: (1) very near-wall concentrations; (2) the downstream changes that occur in PDR due to mixing/dilution and degradation; and (3) the effects of changes in the rheological properties (i.e. MW, intrinsic viscosity, concentration, relaxation time, shear rate for DR onset, critical shear rate for degradation). The authors and collaborators in fact have several papers in the literature that address these topics more completely.

In these experiments (Winkel *et al.*, 2009), three nominal free-stream speeds U were employed (6.65, 13.2, 19.9 ms^{-1}); three PEO nominal MWs corresponding to WSR-N60K, WSR-301, WSR-308 of 2M, 4M, 8M, respectively, were used; three concentrations: 1000, 2000, 4000 wppm were injected; and three volumetric injection rates per unit span, q, of 0.14, 0.28, and 0.71 $ls^{-1}m^{-1}$ were employed. In these flows it is customary though somewhat arbitrary to non-dimensionalize q by dividing by q_s. Here q_s was taken to be 67.3 v, where this represents the usual q in the near-wall region in the buffer layer of the TBL if one assumes a linear velocity profile exists from the solid surface to $y^+ = 11.6$ (Wu and Tulin, 1972). Using this value of q_s gives non-dimensional volumetric injection rates per unit span of 2, 4, and 10 in these experiments.

As the LCC contains about 1.4M gallons (5300 m^3) of water, it was prohibitive to drain the water following each experiment; consequently chlorine was added necessarily to the bulk and monitored to mitigate the build-up of background polymer. As with essentially all of the authors' experimental results, drag measurements had experimental uncertainty of about $\pm 5\%$.

Figure 5 (Figure 1; Tables 1 and 2 of Winkel *et al.*, 2009) presents a schematic drawing of the HIPLATE (name given to this model) with the injection and measurement types and locations evident. On this diagram PIV, LIF, LDV, are the usual acronyms for particle image velocimetry, laser induce fluorescence (or planar laser-induced fluorescence), laser Doppler velocimetry, respectively, and note also the direction of the gravity vector as the working surface of the plate (plate down configuration) is presented for clarity. In addition, the tables in this figure give a summary of the single phase parameters and of the polymer solutions, respectively.

The rheology of polymer solutions remains a timely research topic. The FENE and FENE-P models (as mentioned with regard to Dubief *et al.*, 2004)

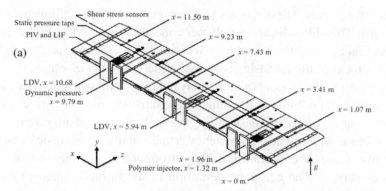

FIGURE 1. Schematic drawing of the large flat-plate test model with injection and measurement locations. The model was inverted in the LCC test section; so the underside of the test model is shown here. The x-axis is parallel to the flow direction.

(b)

	X/L	$X - X_i$ (m)	U (m s^{-1}) $\pm 1\%$	Re_X $\times 10^6$	δ_{99} (mm) $\pm 10\%$	Θ mm $\pm 5\%$	K' $\times 10^{10}$	C_f ± 0.0001	u_τ (m s^{-1})	l_v (μm)	k^+
Injector	0.10	0	6.57	8.7	18	2.2	7.3	0.0024	0.23	4.4	0.1
			13.11	17	17	1.9	3.7	0.0022	0.43	2.3	0.2
			19.75	26	15	2.0	2.4	0.0021	0.63	1.6	0.3
Sensor 1	0.15	0.64	6.58	13	25	2.8	7.3	0.0023	0.22	4.5	0.1
MS1			13.13	26	23	2.6	3.7	0.0021	0.42	2.4	0.2
			19.78	39	21	2.1	2.4	0.0020	0.62	1.6	0.2
Sensor 2	0.26	2.09	6.60	23	38	4.0	7.2	0.0021	0.22	4.6	0.1
			13.18	45	35	3.6	3.6	0.0019	0.41	2.5	0.2
			19.85	67	34	3.6	2.4	0.0018	0.60	1.7	0.2
Sensor 3	0.46	4.62	**6.65**	40	**58**	**5.7**	7.2	0.0020	0.21	4.8	0.1
MS2			**13.25**	79	**55**	**5.3**	3.6	0.0018	0.40	2.5	0.2
			19.93	119	**54**	**5.2**	2.4	0.0017	0.58	1.7	0.2
Sensor 4	0.58	6.11	6.66	50	68	6.5	7.2	0.0019	0.21	4.8	0.1
			13.30	99	65	6.1	3.6	0.0017	0.39	2.5	0.2
			20.03	150	65	6.0	2.4	0.0017	0.58	1.7	0.2
Sensor 5	0.72	7.91	6.69	62	81	7.5	7.2	0.0019	0.20	4.9	0.1
			13.35	124	77	7.1	3.6	0.0017	0.39	2.6	0.2
			20.11	187	78	6.9	2.4	0.0016	0.58	1.7	0.2
Sensor 6	0.83	9.36	**6.70**	72	**90**	**8.2**	7.1	0.0018	0.20	5.0	0.1
MS3			**13.39**	144	**86**	**7.8**	3.6	0.0017	0.39	2.6	0.2
			20.22	217	**88**	**7.5**	2.4	0.0016	0.57	1.8	0.2

TABLE 1. Single phase free-stream velocity U, Reynolds number Re_X, 99 % boundary layer thickness δ_{99}, momentum thickness Θ, acceleration parameter K', skin-friction coefficient C_f, the friction velocity u_τ, the viscous length l_v and the surface roughness parameter k^+ at the downstream locations of the injector, the six skin-friction measurement balances and the three PLIF measurement stations. Bold values result from the measured velocity profiles.

(c)

PEO polymer	M_w (g mol^{-1})	$[\eta]_o$ (cm^3 g^{-1})	c^* (w.p.p.m.)	Δ/C (w.p.p.m.$^{-1}$)	θ_Z (s)	θ_K $/C^{1/2}$ (s)	γ^* (s^{-1})	γ_D (s^{-1})
WSR-N60K	2.3×10^6	1.1×10^3	8.7×10^2	4.4×10^2	4.6×10^{-4}	1.30	1.5×10^3	2.4×10^4
WSR-301	4×10^6	1.8×10^3	5.7×10^2	8.1×10^2	1.2×10^{-3}	2.17	8.4×10^2	7.1×10^3
WSR-308	8×10^6	3.0×10^3	3.3×10^2	1.7×10^3	4.2×10^{-3}	3.58	4.2×10^2	1.5×10^3

TABLE 2. Polymer solution properties.

Figure 5. From Winkel *et al.* (2009): (a) Schematic of the HIPLATE, (b) single phase flow parameters, and (c) polymer solution properties.

are used frequently. These models give reasonable results for dilute polymer solutions. This dilute limit for polymer concentrations is determined from a definition by using the "overlap concentration", c^*, defined as $c^* = [\eta]_o^{-1}$ the reciprocal of the intrinsic viscosity. In the Winkel *et al.* study, $c > c^*$, hence the numerical models would likely not predict these results. Note also that in (c) or their Table 2, the units of the intrinsic viscosity are given as $cm^3 \, g^{-1}$, the reciprocal of mass density while the units of c^* are given in that same table as wppm. The inconsistency is that usually the intrinsic viscosity has units of inverse concentration. To go from one to the other requires the mass density, ρ. The complete definition of the intrinsic frequency (see Elbing *et al.*, 2011) is

$$[\eta]_o = \frac{1}{c} \left(\frac{v_o - v_\infty}{v_\infty} \right) \times 10^6.$$

In this equation, v_o and v_∞ are the kinematic viscosities at zero and infinite shear. Infinite shear viscosity would cause complete scission and thus can be taken as that of the solvent, in this case water. Regardless, one can determine $[\eta]_o$ from the so-called Mark–Houwink relationship (Bailey and Callard, 1959), $[\eta]_o = 0.1248 M_W^{0.78}$. One additional note is that the last five horizontal entries in (c) their Table 2 are computed based on the first three entries (for the equations see Winkel *et al.*, 2009).

Three regions of the flow downstream of the injection of the polymer were described by the authors: (1) Development region; (2) Transition region; and (3) Final region. These will be discussed in-turn.

We begin our in-depth discussion of the results from Winkel *et al.* (2009) with the experimental run that exhibited the highest DR, and as one might guess from our previous discussions, it was the run with the highest MW, 8M, highest concentration, 4000 wppm, and largest injection flux of polymer, $q/q_s = 10$. These results are presented as Figure 6 and include three test speeds, and it can be seen that maximum DRs of 70, 67, and 60% are evident for the lowest to highest free-stream speeds.

First note the shape of the %DR curves for each speed. The shape is related to the length of the three physical regions of the flow, i.e. the length of the Development, Transition, and Final regions. To explain why only the $19.9 \, ms^{-1}$ curve is decreasing monotonically, we digress to discuss the so-called K-factor and the three DR regions.

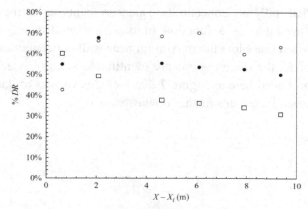

FIGURE 3. The %*DR* versus downstream distance from the injector, $X - X_i$, for WSR-308 for $C_i = 4000$ w.p.p.m. polymer injected at $Q_i/Q_S = 10$ for $U = 6.65$ (○), 13.2 (●) and 19.9 m s^{-1} (□).

Figure 6. Results from Winkel *et al.* (2009) showing the highest levels of DR seen in these experiments.

2.3.2. *K*-factor scaling of PDR

Following Vdovin and Smol'yakov (1981), we review a simple scaling law that is useful for presenting data, although not entirely successful or satisfactory for scaling:

$$K = \frac{q_{inj} c_{inj}}{U(X - X_{inj})} \times 10^{-6}$$

where "*inj*" represents injected and $X - X_{inj}$ is the downstream distance from injection. Physically this scales the injected polymer flux (q_{inj}, c_{inj}) with the free-stream speed and the downstream distance. As demonstrated in Winkel *et al.* (2009) in their Figure 4, the Development region is seen for larger *K*-values (i.e. note that as the downstream distance from the injection point is in the denominator, *K* increases as one moves closer to the injector). For a particular speed, the flux of polymer collapses the data somewhat, indicating that it is a relevant and important grouping. In addition, it was seen that in the Transition region the two higher MW polymers exhibited a logarithmic decrease in DR with decreased *K* and segregated according to molecular weight in the graph of %DR versus *K*. The authors attributed this to likely scission and/or aggregation effects. For additional discussion of the three regions, the interested reader is referred to the paper.

Using PLIF, polymer concentration measurements across the boundary layer were available as a function of downstream distance (see their Figure 7). When one plots the maximum near-wall concentration (c_M) as a function of K, the three regions are identifiable as can be seen in their Figure 9 reproduced here as Figure 7 that includes six sets of data (Recall that K decreasing indicates further downstream.).

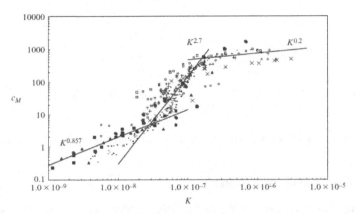

FIGURE 9. The values of c_M versus K for WSR-301 of the present experiment: $U = 6.65$ (●), 13.2 (■) and 19.9 (▲) m s^{-1} with previously reported results; Fruman & Tulin (1976) with $C_i = 100$ w.p.p.m. (○), 500 w.p.p.m. (△) and 1000 w.p.p.m. (□); Vdovin & Smol'yakov (1978), assuming $C_i = 1000$ w.p.p.m. (+); Vdovin & Smol'yakov (1981), assuming $C_i = 1000$ w.p.p.m.; (−); Fontaine *et al.* (1992) (×). The three lines correspond to exponents of 0.2, 2.7 and 0.857. Details of the previous experiments are provided in table 4.

Figure 7. c_M as a function of K taken from Winkel *et al.* (2009).

Clearly, the figure can be divided into three distinct regions as can be seen by the lines superposed on the figure. The rightmost line represents the Development region, the middle line the Transition region, and the leftmost line the Final region. These evolution regimes can be defined as follows.

(1) Development region: The DR increases with distance downstream and the polymer is inhomogeneous and filamentatious.
(2) Transition region: DR begins decreasing as mixing occurs and fewer filaments are present.
(3) Final region: Mixing and dilution occurs through boundary layer growth.

The basic idea to take from this paper then is %DR curves for each of the MW's for each of the polymers. First, the intrinsic %DR is defined as $[\%DR] = \lim_{c \to o}\left(\frac{\%DR}{c_M}\right)$ similar to efficiency, and the intrinsic concentration $[c] = \frac{\%DR_{max}}{[\%DR]}$, both introduced by Virk *et al.* (1967). In addition the following relationship was established:

$$\frac{\%DR}{c_M} = \frac{[c][\%DR]}{[c] + c_M}.$$

Following Little and Patterson (1974), if we take the reciprocal of the above equation $\left(\text{i.e. } \frac{c_M}{\%DR} = \left(\frac{1}{[c][\%DR]}\right)c_M + \frac{1}{[\%DR]}\right)$, the right-hand side of the equation is linear in c_M with slope $\frac{1}{[c][\%DR]}$ and y-intercept $\frac{1}{[\%DR]}$. Fitting a line to the data in this manner yields the intrinsic values, $[c]$ and $[\%DR]$, that are presented in their Table 5 as part of our Figure 8 for each polymer and each speed, and for each polymer without regard to speed. In addition the data points and fitted curves that are presented in their Figure 12 with %DR/c_M as a function of c_M are shown in Figure 8.

Hence for each polymer, and the three speeds tested, knowing c_M, the maximum concentration at the wall, allows you to obtain the %DR from these curves.

2.3.3. The effects of roughness on FDR in a TBL — the experiments of Elbing *et al.* (2011)

Drag reduction, diffusion, and degradation were measured in a TBL with large Re. This was a similar measurement to Winkel *et al.* (2009) in that it included a hydraulically smooth plate; however, a fully-rough surface was tested also. For these experiments only WSR-301 PEO was used. In the near-field, for the slower speeds increased mixing is seen and hence DR; however, downstream a decreased DR was realized in all cases. The roughened surface consisted of tightly packed glass bead (diameters $450 \pm 250\,\mu$m) in Aquapon epoxy paint. The coating was applied over the entire working surface, i.e. the lower surface of the HIPLATE. From $X = 9$ m onward, the surface had increased roughness due to an application error.

Measurements included a single component laser to quantify the free-stream velocity at an upstream reference position, and using pressure sensors, free-stream speeds along the plate were approximated. Skin friction

	U(ms⁻¹)	[%DR](wppm⁻¹)	[c](wppm)
WSR-N60K	6.65	6	7
	13.2	6	8
	19.9	4	12
	All	5	10
WSR-301	6.65	32	2
	13.2	25	2
	19.9	17	3
	All	26	2
WSR-308	6.65	73	1
	13.2	27	3
	19.9	14	5
	All	32	2

Table 5. The intrinsic drag reduction, [%DR], and intrinsic concentration, [c], for the near-wall polymer solutions. The values for each speed and for all speeds combined are presented. The fitted curves are shown in Figure 12.

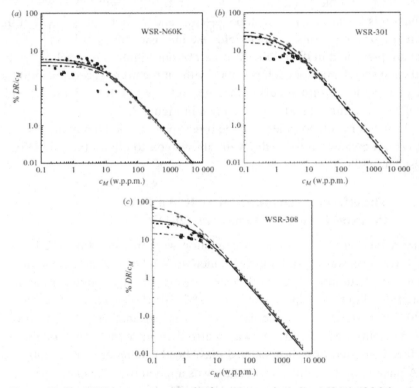

FIGURE 12. The $\%DR/c_M$ versus c_M for the three M_w examined: $U = 6.65$ (○), 13.2 (●) and 19.9 (□) m s⁻¹. Curves fitted to (7.1) are plotted for each speed (dashed lines) and for all speeds together (solid line). Values of intrinsic drag reduction and intrinsic concentration derived from the fitted curves are provided in table 5.

Figure 8. Table 5 and Figure 12 from Winkel *et al.* (2009) to demonstrate the intrinsic drag and concentration from their experiments as well as the fitted curves for their $\%DR/c_M$ as a function of c_M, respectively.

was measured again using the floating plate force balances at six locations. PLIF was used to determine the near-wall concentrations as discussed previously. Sampling of near-wall solutions was made through a specially designed port and then a spectrophotometer was used to measure dye concentration that is proportional to polymer concentration. A Virk-tube apparatus was used along with a cone-and-plate rheometer for the low wppm and higher wppm measurements, respectively. This facilitated the neglect of shear thinning in these experiments.

The usual smooth/polished HIPLATE surface prior to the application of the roughness had a k^+ of O (0.2) with l_v ranging over 1.6–4.9 μm. Following application of the glass beads in the epoxy paint i.e. the rough surface, the k^+ was measured to be O (400) while the l_v spanned 1.1–1.4 μm.

The fully-rough and smooth surfaces results for the single phase case, Elbing *et al.* (2011) Figure 5, is presented in Figure 9. In the same way as any C_D versus Re curves, when C_D ceases to be a function of Re, fully turbulent flow is present, and the fact that sand-grain roughness is a reasonable assumption is borne-out by the following. Using an equation from White (2006) that included the assumption that roughness shifts the log region from the wall, the usual equation becomes

$$u^+ = \frac{1}{\kappa} \ln y^+ + B - \Delta B$$

where for a smooth wall, $\kappa = 0.4$ and $B = 5.2$ (as per White and as discussed earlier). For the rough wall, the shift, ΔB, is obtained from the equation

$$\Delta B = \frac{1}{\kappa} \ln(1 + 0.3 k^+).$$

Figure 6 from Elbing *et al.* (2011) included here as the lower figure of Figure 9 demonstrates the validity of this equation. Hence the conclusion is that the surface is indeed of sand-grain roughness type.

Prior to discussing the rough versus smooth results, for consistency, in Figure 7 of Elbing *et al.* (2011) the experiments from the present smooth surface and WSR-301 PEO is superposed on the data from Winkel *et al.* (2009) along with fitted curves of the data. Although not reproduced here, the data from the two experiments are in reasonable agreement as they should be having used the same HIPLATE model in the same LCC test

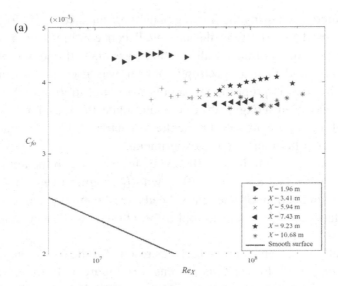

FIGURE 5. Baseline (non-injection) rough-surface skin-friction results for each downstream location at speeds ranging from 6.7 to 20.3 m s^{-1}. The Reynolds number independence at each downstream location indicates that the model was fully rough. Also included for comparison is the smooth-surface best-fit curve.

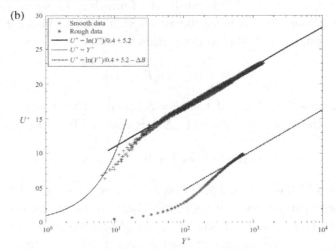

FIGURE 6. Baseline near-wall velocity measurements acquired on the smooth and rough surfaces. The smooth-surface results are from a range of flow speeds (6.6–20.3 m s^{-1}) and downstream distances (1.96–10.7 m), and the rough-surface results are from $X = 1.96$ m at 6.8 m s^{-1}. The dashed line is the predicted log region from the skin-friction balance results, assuming a sand-grain-type roughness.

Figure 9. These figures from Elbing *et al.* (2011) demonstrate that the flow is fully turbulent (a), and that the roughness is of sand-grain type (b).

facility, and conducting the experiment prior to application of the surface roughness.

Next, the polymer injection and DR on smooth versus rough surfaces is addressed. Their Figure 8 is shown as Figure 10.

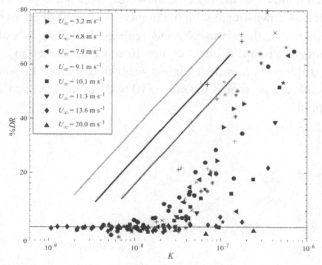

FIGURE 8. Rough-surface %*DR* results as a function of *K*. The smooth-surface best-fit curves from figure 7 are included for comparison. Also included are the results from Petrie *et al.* (2003) of their fully rough surface at 4.6 (+), 7.6 (×) and 10.7 m s^{-1} (*).

Figure 10. Data included on this graph are of three types, and two sets are from Elbing *et al.* (2011) and one set from Petrie *et al.* (2003). The lines shown are fits to smooth surface data (from left to right, lines are for 6.7, 13.4, and 20.1 ms^{-1}) from Elbing *et al.* as are the data points at 3.2, 6.8, 7.9, 9.1, 10.1, 11.3, 13.6, and 20.0 ms^{-1} that are for their roughened surface. The remaining data points at the other speeds (4.6, 7.6, and 10.7 ms^{-1}) are from Petrie *et al.* for their fully rough surface.

The data included in Figure 10 is from Elbing *et al.* and three speeds are from Petrie *et al.* (2003). The lines shown on the figure are fits to smooth surface data (graphed lines from left to right are for 6.7, 13.4, and 20.1 ms^{-1}) from Elbing *et al.* as are the data points for their rough surface for speeds of 3.2, 6.8, 7.9, 9.1, 10.1, 11.3, 13.6, and 20.0 ms^{-1}. The remaining data points at the other speeds (4.6, 7.6, and 10.7 ms^{-1}) are from Petrie *et al.* for their fully rough surface. The surface roughness in the Elbing *et al.* experiment varied over the range $130 < k^+ < 990$. Recall that k^+ varies with l_v which varies with u_*.

As is evident for the low speeds, very close to the injector (K large) the level of DR is increased by the surface roughness in comparison with the hydraulically smooth surface as also found by Petrie *et al.* In general, the rough surface seems to shorten the initial zone, and hence the highest DR is realized closer to the injector; however, what seem positive at first glance is not, as downstream with the higher speeds DR decreases rapidly. In fact at $20\,\mathrm{ms}^{-1}$, there was NO DR realized including at only 56 cm from the injection point! A direct comparison is emphasized in a different manner by Elbing *et al.* in their Figure 9 shown as Figure 11 where three of their speeds are presented for $q/q_s = 10$ with an injected concentration of 4000 wppm.

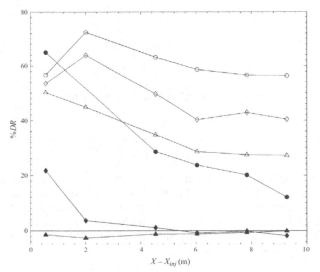

FIGURE 9. The %DR versus the downstream distance from the injection location ($X - X_{inj}$) at a single injection condition ($Q_{inj} = 10\ Q_x$, $C_{inj} = 4000$ w.p.p.m.). Results are from the smooth (open symbols) and rough (solid symbols) surfaces at three nominal test speeds: 6.8 (○), 13.5 (◇) and 20.1 (□)m s^{-1}. Line segments are used to help compare trends between data sets.

Figure 11. Reproduced from Elbing *et al.* (2011), their Figure 9 shows three speeds (6.8, 13.5, and 20.1 ms^{-1}) with a direct comparison of the DR on the rough (solid symbols) and smooth (open symbols) surfaces as a function of distance downstream.

As can be seen in Figure 11, where the open symbols represent the hydraulically smooth surface and the filled symbols the rough surface, the DR realized on the two surfaces differ dramatically. For the rough surface and the lower two speeds, the DR immediately decreases as a function of distance downstream (i.e. the development region is no longer evident for the downstream distance measured), while for the highest speed (20.1 ms^{-1}), to within our $\pm 5\%$ error, no DR is evident. Interestingly for the slowest speed and rough surface, at 56 cm downstream the DR was largest (although not in an absolute sense as the slowest speed on the smooth plate gives the largest overall DR, but this was at the next measurement station — as that flow was still developing at 56 cm). This is due to the rough surface greatly decreasing the development region for polymer injection.

The relative poor performance by the polymer on the rough surface is due to one or more of the following factors:

(1) Enhanced mixing due to higher turbulence.
(2) Effect of roughness on the DR mechanism (i.e. a physical change caused by the roughness).
(3) Polymer degradation increase due to roughness.

In Figure 12 (Figures 10 and 11 from Elbing *et al.*) superposed on graphs presented previously for smooth surfaces, are the results from the rough surface for $c_M = c_M(K)$ and %DR/c_M as a function of c_M. In the first figure note that all the new data lie along the Final zone, $K^{6/7}$, curve implying that the Development and Transition zones may have been decreased or removed/skipped completely. Also, the Final zone is the usual region of TBL growth (that fits the $1/7^{\text{th}}$ power-law velocity profile). The second figure is another clear indicator of reduced DR for the flow over the rough surface.

To identify the culprit causing the reduction in DR more accurately, additional information is required, interested reader is referred to their paper. The bottom line, however, is that most previous studies did not exhibit the level of degradation realized in the LCC facility due to the shorter residence time in their experiments due to the reduced downstream lengths in their physical setups.

2.3.4. Another study on surface roughness effects but conducted at smaller Re

One additional study conducted and published by the authors and collaborators, Elbing *et al.* (2010), is described in this chapter on PDR, although it was

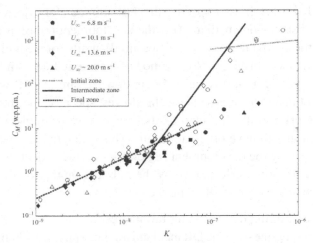

FIGURE 10. The maximum concentration at the wall from the rough surface (solid symbols) versus K. For comparison smooth-surface results (open symbols) reported in Winkel *et al.* (2009) are included. Also plotted are the best-fit curves to the initial, intermediate and final diffusion zones, determined from several research efforts compiled in Winkel *et al.* (2009). Typical measurement uncertainty was ± 50 %, though far downstream it was nearly doubled.

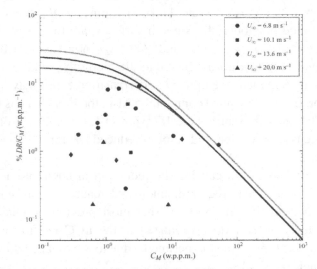

FIGURE 11. Intrinsic drag reduction determined from %DR and maximum concentration measurements on the rough surface ($130 \leqslant k^+ \leqslant 990$). The curves correspond to the WSR301 results obtained by Winkel *et al.* (2009) with a smooth surface at 6.7 (dashed grey), 13.3 (solid black) and 20.1 (dashed black) m s^{-1}.

Figure 12. Figures 10 and 11 reproduced from Elbing *et al.* (2011). The upper figure shows that the solid symbols that represent the new data for the rough surface all lie along the Final zone ($K^{6/7}$). In the lower figure, one can see that the DR is reduced greatly for the flows (over the rough surface) that are again represented by the solid symbols. The curves are representative of the smooth surface results.

published prior to the previously discussed paper. The authors investigated surface roughness effects on a short (0.94 m) plate with speeds to $10.6\,\mathrm{ms}^{-1}$ (Re to 9M) in a 9 in (22.9 cm) water tunnel at the University of Michigan. In this study, the plate surface was hydraulically smooth, transitionally rough (equivalent 240 grit sandpaper), and fully rough (equivalent 60 grit sandpaper). Corresponding k^+ ranges were 0.4–0.7, 13–27, and 74–165, respectively. Using White 2006's ranges for hydraulically smooth ($k^+ < 4$), transitionally rough ($4 < k^+ < 60$), and fully rough ($k^+ > 60$), it can be seen that all flow conditions/surface types were tested. Figure 3 of their paper has been reproduced as Figure 13, and shows the inner variable velocity profiles.

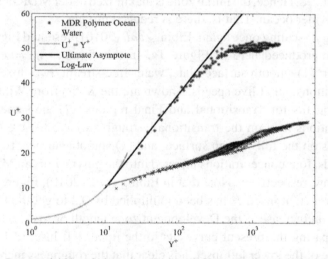

Figure 13. Inner-variable-scaled mean velocity profile as measured at the downstream measurement location in treated water ($\times$) and in a MDR polymer ocean ($*$). The water results were for Re of 10^6 to 10^7; also shown is the log law, $u^+ = y^+$, and Virk's ultimate velocity asymptote. Curves are in good agreement with the appropriate data.

On the figure is the viscous sublayer profile ($u^+ = y^+$), the usual log law (to which much of the non-injected/treated water profiles adhere), and the Virk MDR curve ($u^+ = 11.7\ln(y^+) - 17.0$) with which much of the polymer ocean (PEO WSR-301 with average concentration in the tunnel of

35, a minimum of 30, and a maximum of 45 wppm[1]) data agree. To graph the polymer ocean data against the Virk MDR curve, u_* and l_v are required. Using information based on Larson (2003) that at MDR the FDR is 70% of that for a smooth surface, u_* and l_v were computed. Scaling the Elbing *et al.* (2009) data and graphing it resulted in the figure presented. Note that the data nicely follow the log law and the MDR curve.

Additionally, as we have not mentioned it previously, a comment on the notion of a polymer ocean is warranted. Its purpose was to produce an ideal injection condition necessary to study the initial zone. And, if it is desired to lengthen the initial (Development) zone to provide higher DR for a longer distance, it must be better understood. The idea of an idealized injection condition is based on the fact that during MDR, turbulent eddies are minimized. Hence, the initial zone is maximized under MDR or injection into a polymer ocean. That is, there is reduced mixing.

Using K-scaling once again, Elbing *et al.* (2010) presented Figures 6–9 that are reproduced here as Figure 14. In this figure, $c_M = c_M(K)$ is shown for: (1) smooth surface plate, water free-stream, two fluxes at four concentrations, all at five speeds. Shown are the K-fits from Winkel *et al.* (2009) and fits for Transitional and Final regions; (2) the same for five concentrations now on the transitionally rough surface; (3) the same for six speeds on the fully rough surface; and (4) smooth surface, two fluxes, five speeds, four concentrations injected into the polymer ocean MDR free-stream flow, respectively. Note that in Elbing *et al.* (2010), Figure 9 which is the lower right inset, K has been multiplied by 0.2 to get the fit shown, implying an increase in the Development zone length of five.

Comparing the present curve fits in the upper left inset of Figure 14 with those of the lower left inset, it is clear that the roughness increases the K-value at which the Transitional and Final regions' fits intersect (which corresponds to a shorter intermediate region for the rougher surface). Also visible is that the K-scaling functions is fine for a particular roughness, but differs from one to another, i.e. it is not universal. The most interesting results among the four insets of Figure 14 is the lower right. Recall that

---

[1]Here the exact concentration is only important insofar as it created a polymer ocean condition, and that the authors monitored the conditions to insure this. As the liquid circulated in the tunnel, its molecular weight was in fact decreasing with time due to degradation.

if one can extend the initial region where the DR is very high, the costs associated with PDR decrease, perhaps substantially. The polymer ocean was used to investigate the initial zone, and undeniably it worked toward that end. As can be seen in that inset graph, and is not evident in the three insets that had free-stream treated water, is the Development zone — the seven or eight data points that lie to the high-K side of the Transitional region fitted curve. In this region, the polymer ocean minimizes the turbulence and hence the mixing, and therefore the Development zone is extended as far as possible.

To scale the Development region, Elbing *et al.* followed a parameter of Gebel *et al.* (1978) used in the equation $c_M/c_{inj} = \exp(-X_o/L_o)$ where L_o is the Development length and X_o is the measurement location. If the absolute value of the exponent was less than one, it was assumed within the initial region. This scaling collapsed the data fairly well using l_v and c_M, but λ, the so-called half-diffusion distance required for analysis, was difficult to accurately determine due to the concentration profile; hence Elbing *et al.* used an integral diffusion length (much like that used in turbulence, the integral time scale based on the correlation function of the velocity fluctuations) defined as L^* according to the equation

$$\int_0^{L^*} c(y)dy = \frac{1}{2}\int_0^{\infty} c(y)dy.$$

Thus L^* is the distance from the wall for which the integral of the concentration equals half of the total concentration across the boundary layer. To non-dimensionalize L^*, they used q_{inj}/U, an order of magnitude estimate of the polymer layer thickness at the injector.

Next, the investigators sought the dependence of $(L^*)(q_{inj}/U)^{-1}$ and found that it was a function of (1) the downstream Re, (2) the non-dimensional flux, and (3) the injection concentration. They assumed that the dependence on each of these variables was simply multiplicative with unknown exponents. They further assumed that the manner in which an initial region evolves is similar to a boundary layer or that $L^* \propto \delta$, the boundary layer thickness for the polymer in a MDR ocean. Giles (1968) presented a relationship between the drag coefficient and the Re for the entire plate length. Making several additional assumptions, Elbing *et al.* found that the exponent of Re was 0.7, that of the dimensionless flux (q_{inj}/q_s)

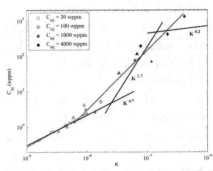

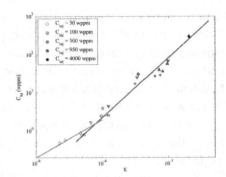

FIG. 6. Average maximum local polymer concentration, C_M, scaled with K for a water background on the smooth surface. Data were collected with injection fluxes between $0.7Q_i$ and $10Q_i$ at free-stream speeds of ($\Diamond$) 5.5, ($\bigcirc$) 7.4, ($\triangle$) 8.5, ($\square$) 9.1, and ($\star$) 10.2 m s^{-1}. The solid lines are best-fit curves to data compiled by Winkel *et al.* (Ref. 22) for the initial, intermediate, and final diffusion zones. The power-law curve fits for the current data in the intermediate (dashed line) and final (gray line) zones are also included.

FIG. 7. Average maximum local polymer concentration, C_M, scaled with K for a water background on the transitionally rough surface. Data were collected with injection fluxes between $2.6Q_i$ and $14Q_i$ at free-stream speeds of ($\Diamond$) 5.6, ($\triangledown$) 6.9, ($\bigcirc$) 7.6, ($\triangle$) 8.6, ($\square$) 9.3, and ($\star$) 10.2 m s^{-1}. The power-law curve fits for the current data in the intermediate (dashed line) and final (gray line) zones are also included.

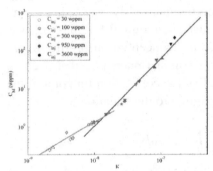

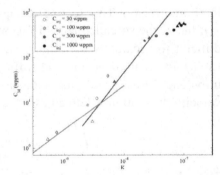

FIG. 8. Average maximum local polymer concentration, C_M, scaled with K with a water background on the fully rough surface. Data were collected with injection fluxes between $1.0Q_i$ and $10Q_i$ at free-stream speeds of ($\Diamond$) 5.7, ($\triangledown$) 7.1, ($\bigcirc$) 7.8, ($\triangle$) 8.8, ($\square$) 9.1, and ($\star$) 10.6 m s^{-1}. The power-law curve fits for the current data in the intermediate (dashed line) and final (gray line) zones are also included.

FIG. 9. Average maximum local polymer concentration, C_M, scaled with K for the MDR polymer ocean background on the smooth surface. Data were collected with injection fluxes between $0.5Q_i$ and $12Q_i$ at free-stream speeds of ($\Diamond$) 5.5, ($\bigcirc$) 7.4, ($\triangle$) 8.5, ($\square$) 9.1, and ($\star$) 10.2 m s^{-1}. The dashed line and gray line are best-fit curves for the intermediate and final zones, respectively, from the smooth surface with a water background (Fig. 6) with K multiplied by 0.2.

Figure 14. The four insets are Figures 6–9 from Elbing *et al.* (2010). In this figure, $c_M = c_M(K)$ is shown for: (1) smooth surface plate, water free-stream, two fluxes at four concentrations, all at five speeds. Shown are the K-fits from Winkel *et al.* (2009) and fits for Transitional and Final regions; (2) the same for five concentrations now on the transitionally rough surface; (3) the same for six speeds on the fully rough surface; and (4) smooth surface, two fluxes, five speeds, four concentrations injected into the polymer ocean MDR free-stream flow, respectively.

was -0.74, and that of the injection concentration, c_{inj}, was 0.142. The interested reader is referred to their paper for these additional assumptions and derivations. Using this scaling produced Figure 15 that was lifted from their Figure 10.

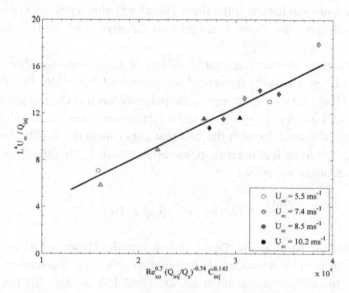

FIG. 10. Integral diffusion length scaling determined from a smooth surface with a MDR polymer ocean. Data collected at nominal injection concentrations of ($\bigcirc$) 30, ($\square$) 100, ($\triangle$) 300, and ($\diamond$) 1000 wppm. The solid line is the linear best-fit curve to the data.

Figure 15. Scaling of the integral diffusion length determined from a smooth surface with a MDR polymer ocean (Elbing *et al.*, 2010).

Figure 15 provides an engineer with the downstream growth of the polymer boundary layer (δ_p) as it goes like L^* as a function of speed and downstream distance, injection flux, and injection concentration, all in the Development region or

$$L^* = (4.16 \times 10^{-4}) \frac{X_o^{0.7}}{\upsilon U^{0.3}} q_{inj}^{0.26} q_s^{0.74} c_{inj}^{0.142}.$$

This shows that the diffusion integral length scale, L^*, increases with downstream distance, X_o, increases with increased volumetric flux of polymer, q_{inj}, decreases with increased speed, U, and increases with increased injection concentration, c_{inj}. The latter finding is somewhat surprising and implies that using large-concentration polymer increases the diffusion (L^* increases) and so polymer is not retained in the near-wall. This is possibly due to the occurrence of filamentation and subsequent three-dimensional nature of the flow. Therefore using a polymer ocean, i.e. ideal conditions, has allowed insight into the physics of the Development region.

If we now choose reasonable values of the parameters for surface ships — $U \sim 10\,ms^{-1}$; numerical integration of the MDR line through $y^+ = 30$ (beyond the buffer region the polymer has no effect as per Dubief *et al.*, 2004) gives $q_{inj} \sim 6\,q_s$ so to be conservative say $q_{inj}/q_s = 10$; the fitted line of Figure 15 with the fact that experimentally for the ordinate, L^*U/q_{inj}, the fitted line maximum was about 15; and using the Gebel *et al.* (1978) equation, we obtain

$$\frac{L_o U}{\upsilon} = (2.33 \times 10^6)(c_{inj} \times 10^{-6})^{-0.2}.$$

a relationship for L_o, the Development length. Hence to increase L_o, one should inject low concentration polymer mixtures. Remember that in addition the concentration must be such that DR is high. To make this occur, then requires both L_o and %DR as a function of c_{inj}.

The latter is available from a relationship and data discussed previously (Figure 12 of Winkel *et al.*, 2009). For WSR-301 PEO polymer, nominal values are chosen here: [c] = 2.5 wppm, [%DR] = 28 wppm^{-1}. Using $X_o = L_o$ in the Gebel *et al.* relationship provides $c_M = c_M(c_{inj})$. And, using the above equation for L_o, the following are graphed in Figure 16: %DR, L_o, and their product, where the latter term is an efficiency of sorts.

These data then suggest that the maximum Development length that one could expect even under ideal conditions is about 1.06 m with a $c_{inj} = 30$ wppm. In reality, L_o would be less and hence on a surface ship, the effect of maximizing L_o would be insignificant.

The intermediate or Transitional zone is wherein the filaments occur and then are diffused across the boundary layer. Although used previously with non-universal results, Elbing *et al.* used curve fitting of a power law to interpret the intermediate zone. As demonstrated in Figure 14, the K-relationship worked for each roughness, but failed for the set. As the

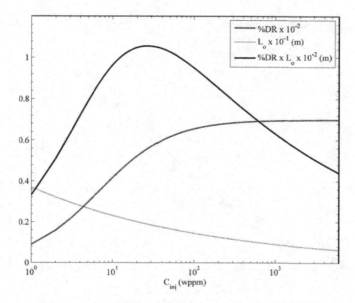

FIG. 11. The maximum initial zone length and corresponding %DR vs injection concentration determined from MDR polymer ocean results. Also included is the product of the initial zone length and corresponding %DR, which illustrates the optimized initial zone length is on the order of one meter.

Figure 16. Shown is the maximum Development zone length along with %DR, L_o, and their product as a function of c_{inj}. These results are for a polymer ocean of MDR concentration, and for reasonable ship parameters. As can be seen in the figure, the maximum Development length is about 1 m.

Development region has a significant effect on the Transition region and the former is sensitive to roughness, k, and the diffusion in the intermediate zone is necessarily a function of the turbulence there, the authors assumed that $k^+ = k/l_v = k/(v/u_*)$ might well supplement K for the second region. Using such an expression, they found that $c_M = c_M(K, k^+) = (K)(1 + k^+)^{-0.17}$, and that fit is presented in Figure 17. A reasonable fit can be seen for all five concentrations tested, i.e. the data scale well.

At this juncture in the research, it is judicious to say that the authors had successfully scaled the Development and the Transition zones across the range of roughness. The Final zone scaling, the best understood of the three, showed that $c_M = c_M(K^{6/7})$ as would be expected.

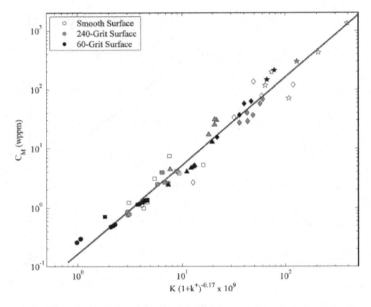

FIG. 12. Intermediate zone scaling of data from the smooth, transitionally rough, and fully rough surface conditions at nominal injection concentrations of (○) 30, (□) 100, (△) 300, (◇) 1000, and (☆) 4000 wppm. The average roughness height was scaled with the baseline viscous wall unit.

Figure 17. Transition region scaling from the three surface finishes: smooth, transitionally rough, and fully rough. Injected concentrations were 30, 100, 300, 1000, and 4000 wppm.

2.4. Polymer Release with Coatings

One final technique that has not been discussed, but warrants mention regarding PDR is that of controlling the rate at which, for example, solid PEO dissolves to facilitate its use as a coating on objects for which DR is desired. And, as the object moves through the water, the solid dissolves and "feeds" the liquid flow with polymer. Several years ago this idea was proposed to funding agencies by one of the authors (MP), but was not supported. The idea is that by controlling the manufacturing process (i.e. the crystallization of PEO), the rate at which the polymer dissolves in water can be altered over a large range, and then knowing the speed of transit of the object and the ppm required to reduce drag appropriately, an appropriate coating can be designed.

Before this chapter is concluded, one very recent publication that used the addition of powdered polymer particles to self-polishing marine paints is discussed. The notion and results are from a paper by Yang *et al.* (2014) of Pusan National University, with the present authors in a supporting role. The idea behind the method is straightforward: mix powdered PEO with self-polishing copolymer anti-fouling (marine) paint to produce DR as the surface reacts with (sea)water in its usual manner. This avoids the issues associated with injecting polymers through the hull. In experiments to assess the technique, reduction of skin friction was as large as 30–40% in a water tunnel as compared to the rough surface left by anti-fouling paint itself and about 10% for total DR in a tow tank. The overarching question is not whether the coating works, but how long it can be made to last. Overall, five different types of experiments were conducted.

As with self-polishing copolymer anti-fouling paints that release biocides (chemicals that kill living organisms) by hydrolysis ion exchange of the paint in water, in this case the PEO (MW = 4M) is released concurrently. The PEO particles added by the researchers had a mean size of 70 microns, and the method of producing them is discussed in the paper.

To investigate the release rate of the PEO, a Taylor–Couette-flow type apparatus was used. In this experiment, rotational motion is used to simulate the conditions of shear in an external flow (i.e. where the freestream contains no polymers) and hence polymer-laden liquid needs to be removed from the annular region and replaced with fresh liquid. Here the rotating inner cylinder had a diameter of 320 mm, the fixed outer cylinder inside diameter was 330 mm, the lengths of the two cylinders were 300 mm and 320 mm, respectively, and the purging of the liquid proceeded at 1 ml s^{-1}. Figure 6 of their paper is reproduced as Figure 18, and shows the PEO concentration as a function of time. Although the polymer release rate drops precipitously at first, both the PEO1% and the PEO2% (where the 1% and 2% denote mass fraction in the paint mixture, respectively) achieve an asymptotic state prior to the end of the test of 50 h. Note that the ordinate values, which are PEO concentrations in mg l^{-1}, are very close to wppm as the specific gravity of the solution is approximately one.

The second set of experiments used to explore the method of polymer release (i.e. "injection") from the anti-fouling paint was low-Re-number tests in a water tunnel. In this investigation, four individual surfaces

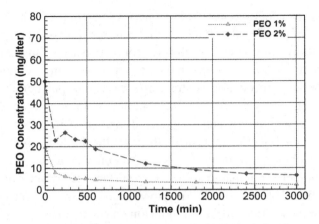

Figure 18. Reproduction of Figure 6 from Yang *et al.* (2014) depicting the temporal change of the PEO concentration measured in the annular region of the flow apparatus.

were installed that included uncoated, hydraulically-smooth aluminum; AF (anti-fouling) paint used as a baseline; a PEO1% coating; and a PEO2% coating, both as defined above. It is emphasized strongly that the primary purpose of these marine paint coatings is for use on ships, and as such the comparisons were made primarily to the AF paint coatings, not to the historical smooth surface results. Moreover, as the paint coatings are quite rough, the DR of the PEO polymer-laden paints are large compared to the AF surfaces, but less when compared to the smooth surfaces. Relative to the chosen baseline, for the speed tested, the %DR was about 12 and 17% for the two coatings as can be seen in their Table 1 reproduced here as Table 1.

Table 1. Reproduction of Table 1 from Yang *et al.* showing the %DR for each of the surfaces tested in the low Re number experiments.

Sample	U_M (m/s)	Re_M	$C_f \times 10^3$	%DR (Percentage Drag Reduction) Relative to smooth	Relative to baseline AF
Smooth (Uncoated)	0.535	28,200	4.471	—	—
Baseline AF	0.535	31,800	4.932	−10.3	—
PEO1%	0.535	32,700	4.355	2.59	11.7
PEO2%	0.536	32,800	4.090	8.52	17.1

In the third and fourth sets of experiments conducted in a tow tank, flat plates and two identical ship models were towed. For the ship models, power saved overall for the PEO1% as compared to the AF surface was about 5%. Lastly, experiments on friction reduction for flows along coated coupons were tested in the facility at the University of Michigan at higher Re [to $O(10^7)$] and reductions of to 40% were realized. It is noted that this notion is very new, and additional investigation is warranted.

This concludes the chapter on PDR. Much can be gleaned from the discussions on how to design a system and the success you might expect. The bottom line is that PDR can work under certain conditions where the turbulence is such as not to degrade the chains, and as we will see later, if the cost of injection is outweighed by another objective such as increased speed over short distances such as might be desirable for defense purposes. However, PDR is not practical economically for long-haul conditions at high speeds as we will see in our economic analysis that will be presented in Chapter 4.

References

Bailey, F.E. and Callard, R.W. (1959). Some properties of poly(ethylene oxide) in aqueous solutions. *J. Appl. Polym. Sci.* 1(1), 56–62.

Berman, N.S. (1978). Drag reduction by polymers. *Annu. Rev. Fluid Mech.* 10, 47–64.

Boyer, R.F. and Miller, R.L. (1977). Polymer-chain stiffness parameter, sigma, and cross-sectional area per chain. *Macromolecules* 10(5), 1167–1169.

Dubief, Y., White, C.M., Terrapon, V.E., Shaqfeh, E.S.G., Moin, P. and Lele, S.K. (2004). On the coherent drag-reducing and turbulence-enhancing behaviour of polymers in wall flows. *J. Fluid Mech.* 514, 271–280.

Elbing, B.R., Winkel, E.S., Solomon, M.J. and Ceccio, S.L. (2009). Degradation of homogeneous polymer solutions in high shear turbulent pipe flow. *Exp. Fluids* 47, 1033–1044.

Elbing, B.R., Dowling, D.R., Perlin, M. and Ceccio, S.L. (2010). Diffusion of drag-reducing polymer solutions within a rough-walled turbulent boundary layer. *Physics of Fluids* 22(4), 045102, 1–13.

Elbing, B.R., Solomon, M.J., Perlin, M., Dowling, D.R. and Ceccio, S.L. (2011). Flow-induced degradation of drag-reducing polymer solutions within a high-Reynolds number turbulent boundary layer. *J. Fluid Mech.* 670, 337–364.

Garwood, G.C., Winkel, E.S., Vanapalli, S., Elbing, B., Walker, D.T., Ceccio, S.L., Perlin, M. and Solomon, M.J. (2005). Drag Reduction by a Homogenous Polymer Solution in Large Diameter, High Shear Pipe Flow. ISDR, Busan, Korea.

Gebel. C., Reitzer, H. and Bues, M. (1978). Diffusion of macromolecular solutions in the boundary layer, *Rheol. Acta* 17, 172.

Hoyt, J.W. (1972). Effects of additives on fluid friction. *J. Basic Eng.* 94(2), 258–285.

Kundu, P.K., Cohen, I.R. and Dowling, D.R. (2012). *Fluid Mechanics* Academic Press Elsevier, Amsterdam.

Larson, R.G. (1999). The structure and rheology of complex fluids. Oxford University Press, New York.

Liaw, G.C., Zakin, J.L. and Patterson, G.K. (1971). Effects of molecular characteristics of polymers on drag reduction. *AICHE J* 17(2), 391–397.

Little, R.C. and Patterson, R.L. (1974). Turbulent friction reduction by aqueous poly(ethylene oxide) polymer solutions as a function of salt concentration. *J. Appl. Polym. Sci.* 18, 1529–1539.

Lumley, J.L. (1969). Drag reduction by additives. *Annu. Rev. Fluid Mech.* 1, 367–387.

McComb, W. (1990). The physics of fluid turbulence. Oxford University Press, Oxford.

Nieuwstadt, F. and Den Toonder, J. (2001). Drag reduction by additives: A review. In: Soldati, A. and Monti, R. (Eds.), *Turbulence Structure and Motion*, pp. 269–316. Springer, New York.

Oweis, G.F., Winkel, E.S., Cutbirth, J.M., Perlin, M., Ceccio, S.L. and Dowling, D.R. (2010). The mean velocity profile of a smooth flat-plate turbulent boundary layer at high Reynolds number. *J. Fluid Mech.* 665, 357–381.

Patterson, R.W. and Abernathy, F.H. (1970). Turbulent flow drag reduction and degradation with dilute polymer solutions. *J. Fluid Mech.* 43(4), 689–710.

Petrie, H.L., Brungart, T.A. and Fontaine, A.A. (1996). Drag reduction on a flat plate at high Reynolds number with slot-injected polymer solutions. In: *Proceedings of the ASME Fluids Engineering Division* ASME, New York, 237, 3–9.

Sellin, R.H.J., Hoyt, J.W., Pollert, J. and Scrivener, O. (1982). The effect of drag reducing additives on fluid flows and their industrial applications: Part II. Basic applications and future proposals. *J. Hydraul. Res.* 20(3), 235–292.

Toms, B.A. (1948). Some observations on the flow of linear polymer solutions through straight tubes at large Reynolds numbers. *Proc. First Int. Congr. Rheol.* 2, 135–141.

Vanapalli, S.A., Ceccio, S.L. and Solomon, M.J. (2006). Universal scaling for polymer chain scission in turbulence. *Proc. Natl. Acad. Sci. USA* 103(45), 16660–16665.

Vanapalli, S.A., Islam, M.T. and Solomon, M.J. (2005). Scission-induced bounds on maximum polymer drag reduction in turbulent flows. *Phys. Fluids* 17, 095108-1–095108-11.

Vdovin, A.V. and Smol'yakov, A.V. (1981). Turbulent diffusion of polymers in a boundary layer. *J. Appl. Mech. Tech. Phys.* 22, 98.

Virk, P.S. (1975). Drag reduction fundamentals. *AICHE J* 21(4), 625–656.

Virk, P.S., Merrill, E.W., Mickley, H.S., Smith, K.A. and Mollo-Christensen, E.L. (1967). The Toms phenomenon: Turbulent pipe flow of dilute polymer solutions. *J. Fluid Mech.* 20(2), 305–328.

White, F.M. (2006). *Viscous Fluid Flow*. 3rd edition. McGraw-Hill, New York.

White, C.M. and Mungal, M.G. (2008). Mechanics and prediction of turbulent drag reduction with polymer additives. *Annu. Rev. Fluid Mech.* 40, 235–256.

Winkel, E.S., Oweis, G.F., Vanapalli, S.A., Dowling, D.R., Perlin, M., Solomon, M.J. and Ceccio, S.L. (2009). High Reynolds number turbulent boundary layer friction drag reduction from wall-injected polymer solutions. *J. Fluid Mech.* 621, 259–288.

Wu, J. and Tulin, M.P. (1972). Drag reduction by ejecting additive solutions into pure-water boundary layer. *J. Basic Eng.* 94, 749.

Yang, J.W., Park, H., Chun, H.H., Ceccio, S.L., Perlin, M. and Lee, I. (2014). Development and performance at high Reynolds number of a skin-friction reducing marine paint using polymer additives. *Ocean Engineering* 84, 183–193.

Chapter 3

Friction Drag Reduction by Bubble/Gas Injection

As there has been a recent *Annual Review of Fluid Mechanics* review of this topic by one of the authors (Ceccio, 2010), the interested reader is referred there. Much of that publication discusses cavities and air layers for underwater and surface vehicles — these discussions follow later in this text. To begin our discussion, we mention the first paper that discusses successful drag reduction by bubble injection, McCormick and Bhattacharyya (1973). For a review of bubble drag reduction (BDR) through 1992, the reader is directed to a review by two Pennsylvania State University faculty, Merkle and Deutsch (1992). Here, we reproduce an introductory paragraph from the paper Sanders *et al.* (2006) that provides some general background.

> A comprehensive review of the first two decades of work on BDR is provided by Merkle and Deutsch (1992); hence, summary provided in the following paragraphs is brief and emphasizes flat-plate studies. McCormick and Bhattacharyya (1973) reported the earliest successful BDR experiments (Re_x to 1.8×10^6). About the same time, Soviet workers (Migirenko and Evseev, 1974; Bogdevich and Evseev, 1976; Bogdevich and Malyuga, 1976) observed that the maximum level of drag reduction occurred immediately downstream of gas injection and did not persist with increasing downstream distance. Madavan *et al.* (1984) made BDR measurements in a zero-pressure-gradient turbulent boundary layer at Re_x as large as 10^7 and found that BDR improved with decreasing flow speed, increasing gas injection rates, and when buoyancy pushed bubbles toward the plate surface. Madavan *et al.* (1985) also determined that the pore size of the air injector had little effect on BDR for pore sizes from 0.5 to 100 μm. Pal *et al.* (1988) found that bubbles must be within 200 wall units of the surface to produce noticeable BDR. Fontaine and Deutsch (1992) determined that BDR did not depend on the density

or composition of the injected gas. Takahashi *et al.* (2001) examined BDR at Re_x as high as 25 million and found that variations in bubble size and boundary-layer thickness did not significantly influence the level of drag reduction. Kawamura *et al.* (2003) found a similar insensitivity to bubble size for bubble radii of 250 to 1000 μm at similar Re_x values. BDR studies have also involved axisymmetric bodies (Deutsch and Castano, 1986; Clark and Deutsch, 1991) and sea trials (Kodama *et al.*, 2000).

The purpose of the Sanders *et al.* investigation was to reveal basic mechanisms associated with BDR and to evaluate BDR downstream persistence. For BDR to be an effective economic drag reduction technique, downstream persistence or simply persistence is a necessity. And by persistence, we mean the ability of the bubbles to remain in the very near-wall region where they cause drag reduction (Other work to be discussed that has been conducted by the group from the University of Michigan, the authors and their collaborators, will include (1) bubble-size effects; (2) saltwater and surfactant effects; (3) and the very important topic of how bubbly layers and flows transition to air layers, the topic of the next chapter.).

The primary parameters in BDR are: (1) gas injection volumetric flow rate per unit span; (2) free-stream speed of the flow; (3) boundary layer state (e.g. turbulence level and thickness) at the location of gas injection; and (4) the distance in the flow direction from the injection location. To collapse data from bubble injection experiments, various scalings have been employed. Similar to the K-scaling used in polymer injection, researchers have used Q_a, the volumetric injection rate, divided by $U \times S$, the product of the free-stream speed and S, a representative area. If $S = l \times X$ with l, the injector span and X, a downstream distance, then $Q_a/l \equiv q_a$ is a volumetric flow rate per unit span and the expression is similar to $q_{inj} \times c_{inj}$ for polymers with the denominator left with $U \times X$ similar to $(U) (X - X_{inj})$. Improved data collapse was seen for a boundary layer void-fraction flux $Q_a/(Q_a + Q_w)$ with Q_w the volumetric flow rate in the absence of injection ($Q_w = (\delta - \delta^*)U\,b$ with the usual notation for the boundary layer thickness and the displacement thickness, and b the span) as shown by Madavan *et al.* (1985). These scalings have met with limited success.

As discussed in Ceccio (2010) and Sanders *et al.* (2006), a significant amount of numerical and analytical work has been conducted, but as yet, no general scaling has been found. Much of the confusion and difficulty in trying to scale the results across different experiments points to a variety of physical processes that are concurrently affecting the experiments, both physical and numerical. As is the case for PDR, numerical simulations suggest that as bubbles can modify the turbulent fluctuations, they work best when they are the size of the order of the smallest (Kolmogorov) turbulence scales. Assuming that this is correct, this criterion presents a difficult hurdle as extremely small size bubbles can be produced (e.g. via electrolysis), but not in sufficient quantities for actual drag reduction cases of interest (As will be discussed subsequently, the authors and co-workers have tried several methods of producing very small bubbles including: (1) coating the bubbles by having lipids present and using sonic techniques; (2) electrolysis that requires that the wire diameter is of the same order of magnitude as the size of the bubble required — at these small diameters the wires are extremely fragile — and it is difficult to produce them in quantity; and (3) others have tried venturi effects for example.). Note that the mechanism suggests as it did with polymers that inner variables of the boundary layer are likely required for scaling.

Again as was the case with PDR, much of the authors' BDR research was conducted at the Large Cavitation Channel of the U.S. Navy. For a general discussion of this facility and its capabilities, see Etter *et al.* (2005). As before, the measurement stations were typically six in total with nominal free-stream speeds of 6^+, 12^+, and 18^+ ms^{-1}. These experiments were conducted with the hydraulically smooth plate as evidenced by the k^+ values all being less than or equal to 0.2 and with the roughened plate.

In the Sanders *et al.* experiment, compressed air was used as the working gas and injected at $X = 1.32, 9.79$ m and where these locations were referred to as upstream (UI) and downstream (DI) injection, respectively. Injection was conducted at four flow rates ($Q_a = (q_a)(2.65) = 0.05, 0.09, 0.19, 0.38$ m^{3}s^{-1}). The injector used for these experiments is shown below (see Figure 19) and included a 40 μm sintered section across the slot, although subsequent experiments demonstrated that the sintered material was not necessary. Later experiments also showed that results were insensitive to the injector angle for the angles tested — from $6°$–$25°$ (Parenthetically, we mention that for most applications of BDR and ALDR (air layer drag reduction), air compressors may not be necessary as high volume, low pressure blowers may suffice and likely are less expensive to operate.).

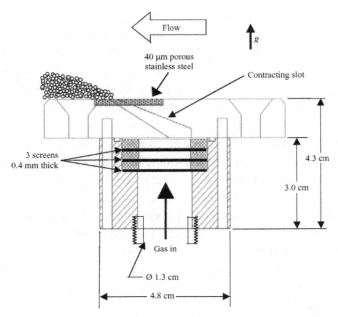

FIGURE 3. Cross-sectional schematic of the gas injectors. The injector measures 2.65 m in the spanwise direction. The gas inlet at the bottom of the injector consists of forty 1.3 cm ports spaced evenly across the span of the injector. Three brass perforated plates with 0.5 mm diameter holes serve to distribute the gas evenly across the injector. The contracting slot has a 10° taper and a 25° mean angle leading to the 40 μm pore-size sintered stainless steel strip.

Figure 19. Diagram of the injector used in the Sanders *et al.* (2006) experiments. It turned out that the 25° mean angle of the 10° contracting slot that led to 40 micron sintered material was not necessary. Even in the absence of the sintered material and with slots ranging from 6° through 25° bubbles and air layers were generated easily.

The instrument suite for this series of experiments included: (1) LDV to measure U; (2) static pressure transducers; (3) gas mass-flow sensors to measure Q_a; (4) skin-friction force plates; and (5) bubble viewing systems within the plate looking at the flow immediately adjacent to the solid surface and outside the tunnel viewing the flow from afar. These latter instruments were used to obtain area ratios rather than true void fractions, γ. A systematic bias estimate existed in the data; however, when the ratio of local skin friction bias was computed, this error was mitigated (In follow-on experiments, the source of the bias error was eliminated.).

The first data that we examine from these measurements are the ratios of the local skin friction coefficients with injection, C_F, to those of the single phase flow (i.e. those without injection, C_{Fo}, as a function of the various downstream measurement stations for the different free-stream speeds and gas injection rates). C_F and C_{Fo} are defined as usual as the ratio of the shear stress divided by the dynamic pressure, equivalent to the shear force divided by the product of the dynamic pressure and the area over which the pressure stress acts. These data are shown in Figure 20 (Figure 6 of Sanders *et al.*) This is a very interesting figure indeed as it foreshadows much of our discussions to follow.

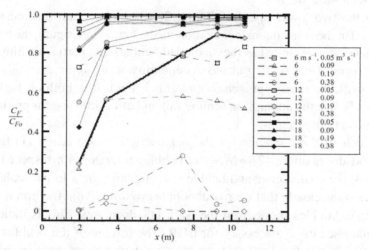

FIGURE 6. Skin friction ratio C_F/C_{Fo} as a function of downstream distance x (in metres) for upstream injection. Error bars are not shown for clarity but would be $\pm 10\%$ to $\pm 20\%$ of each measurement. Data points connected by solid lines represent flow conditions having distinct bubbles. Data points connected by dashed lines represent flow conditions leading to a continuous or intermittent gas film on the surface of the test model with only occasional patches of bubbles. The $12\,\mathrm{m\,s^{-1}}$ and $0.38\,\mathrm{m^3\,s^{-1}}$ test condition shown with a dotted line produced both bubbles and gas film in nearly equal amounts. Measured values of C_F/C_{Fo} that fell below $C_F/C_{Fo} = 0$ because of experimental errors have been plotted at $C_F/C_{Fo} = 0$. The thin vertical line at the left-hand edge of the figure marks the location of the upstream injector.

Figure 20. Graph of the ratio of skin friction coefficients for the UI injection case as a function of distance downstream on the plate. Each of the three nominal speeds, 6, 12, and $18\,\mathrm{ms^{-1}}$ are presented with four injection volumetric flow rates.

What are the most obvious implications of the data presented in Figure 20?

(1) As Q_a decreases for any particular speed, so does DR.
(2) For the lowest speed and largest injection flux, $C_F/C_{Fo} \to 0$.

(3) For many cases, the largest DR is lost rapidly downstream.
(4) At $X = 10\,\text{m}$, 7 of 12 cases along with all of the $18\,\text{ms}^{-1}$ cases have lost their DR.

Of course, scaling and/or non-dimensionalizing these data is desirable and fundamental. Using the Madavan *et al.* (1985) scaling, some collapse of the Sanders *et al.* data was evident, see the first inset of Figure 21, but the other researchers experiments exhibited much more scatter and did not scale with these data.

In the two lower insets of Figure 21, the momentum thickness was included in the non-dimensionalization. In the lower right figure, the $6\,\text{ms}^{-1}$ data have been removed as they included bubbles and partial air film. Due to the form of the scaling abscissa denominator, as q_o or $q_o - q_{o,inj}$ get very small, the ratio approaches one. This is evident in both of the lower figures. Hence the attempts at scaling fail and offer little use for prediction purposes at other scales.

Much of the remainder of the manuscript is focused on (1) bubble size and distribution, (2) where the bubbles migrate with respect to the wall, (3) the coalescence and bubble splitting, and such. In particular, the authors show clearly that near-wall bubble proximity/void fraction is a key factor. As bubbles move from the wall, DR decreases. This is indicative that inner scaling is necessary for BDR. The forcing of the bubbles from the wall when in fact buoyancy was forcing them toward the working (i.e. underside) surface of the model is the next topic of discussion. We include a more complete discussion than was given originally in Sanders *et al.* (2006) by including work from Sanders (2004) and Elbing *et al.* (2013).

We begin with a single, fixed diameter bubble in bastardized flow and investigate the forces it is under in a turbulent boundary layer profile, but with mean liquid velocity vector, $\bar{u}_f$. Although this is far from groups of bubbles in turbulent flow, it turns out that the model captures much of the relevant physics. Maxey and Riley (1983) contributed and Magnaudet and Eames (2000) gave the following equation of motion for a particle

$$\rho_b V_b \frac{d(u_b)}{dt} = F_1 + F_A + F_D + F_L + F_B + F_p$$

where ρ_b and V_b are the mass density and fixed volume of the bubble, u_b is the bubble velocity vector (u_b, v_b), and the right-hand side forces are

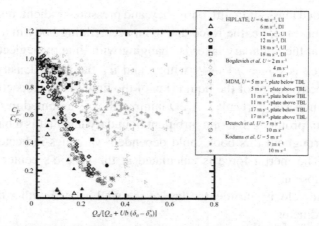

FIGURE 7. Skin friction ratio as a function of the volumetric fraction of gas flow rate (1.2) for the present work, compared to the results of previous researchers. UI = upstream air injection at $x = 1.32$ m; DI = downstream air injection at $x = 9.79$ m. The data from Bogdevich *et al.* represent those acquired by Soviet researchers (1974–1976) in plate-up and plate-down experiments. MDM, data reported by Madavan, Deutsch & Merkle (1984, 1985*a*).

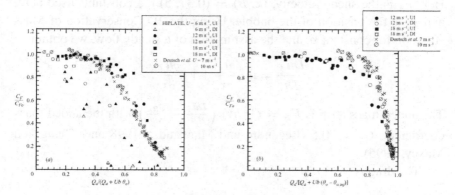

FIGURE 8. (*a*) Skin friction ratio as a function of flux-based void fraction using the momentum thickness of the unmodified boundary layer, θ_o. (*b*) Skin friction ratio as a function of the rescaled gas flow rate, where $\theta_{o,inj}$ is the estimated momentum thickness of the unmodified boundary layer at the streamwise location of the gas injector. The smooth-wall data presented by Deutsch *et al.* (2003) are also plotted.

Figure 21. Figures 7 and 8 from Sanders *et al.* (2006) showing C_F/C_{Fo} as a function of various abscissa non-dimensional quantities that include displacement thickness and momentum thickness.

inertia, added mass, drag, lift, buoyancy, and pressure gradient, respectively
(The Basset force, i.e. the force due to an accelerating body in a fluid and
the fact that the boundary layer is changing with time, is neglected.). The
inertia force is $F_I = \rho V_b \frac{D\bar{u}_f}{Dt}$ with ρ and $\bar{u}_f$ the mass density and the
mean velocity vector of the liquid (i.e. of the flow field), respectively, and
D/Dt the material derivative. A 2D laminar flow is assumed in (x, y), but
the velocity profile is that of a turbulent boundary layer (It is a bastardized
flow.). Throughout this book, bold dependent variables indicate a vector
quantity. The inertia force is calculated at the sphere's center as if the
sphere is absent.

For the velocity profile, the law of the wall is used with x the along-
plate coordinate:

$$\bar{u}_f = \frac{u_*}{\kappa} \ln \frac{y_b u_*}{\upsilon} + u_* B.$$

Here u_* is the shear velocity, $(\kappa, B) = (0.37, 5)$ the constants used here,
and y_b is the position of the bubble. Using the 2D Conservation of Mass
equation for constant ρ, and the assumption of a steady flow, we obtain

$$\frac{D\bar{u}_f}{Dt} = \bar{u}_f \frac{\partial \bar{u}_f}{\partial x} + \bar{v}_f \frac{\partial \bar{u}_f}{\partial x}.$$

The added mass term is $F_A = C_m \rho V_b \left(\frac{D\bar{u}_f}{Dt} - \frac{du_b}{dt} \right)$ with the added mass
coefficient, $C_m = 0.5$ (Legendre and Magnaudet, 1998 and Chang and
Maxey, 1995).

The drag is

$$F_D = \frac{1}{2} C_D \rho A_b |\bar{u}_f - u_b|(\bar{u}_f - u_b)$$

where $C_D = 24\text{Re}_b^{-1}(1 + 0.197\text{Re}_b^{0.63} + (2.6 \times 10^{-4}\text{Re}_b^{1.38})$ and $\text{Re}_b = \frac{2R|\bar{u}_f - u_b|}{\upsilon}$ with R the bubble radius and A_b the projected area of the bubble
(Haberman and Morton, 1953).

The lift force is

$$F_L = C_L \rho V_b (\bar{u}_f - u_b) \times (\nabla \times \bar{u}_f)$$

where the last term in the parentheses represent the vorticity in the flow that
would have been present at the center of the bubble had it not been there.

As with C_m above, a constant value of 0.5 is taken for C_L (Magnaudet and Eames, 2000).

The buoyancy force is simply $\boldsymbol{F}_B = -\rho V_b \boldsymbol{g}$ with V_b the volume of the bubble.

The pressure gradient force is simply $\boldsymbol{F}_p = -V_b \nabla P$ where P is the static pressure.

Other assumptions required include a clean and small bubble with constant spherical volume (i.e. $We \ll 1$, where We is the Weber number).

Using these equations, Sanders solved for $\boldsymbol{u}_b$ and obtained (x_b, y_b) as a function of time. Other than in her thesis, the trajectories presented below have not been published; however this simple model does provide much of the relevant physics.

Figure 47a reproduced here from Sanders (2004) as Figure 22 exhibits the bubble paths begun at the plate surface 1 m downstream from the origin with $U = 18\,\text{ms}^{-1}$ and gravity neglected for four bubble radii. Both the abscissa and the ordinate have been scaled with inner variables. Taking the ratio of the vertical components of the lift to drag give

$$\frac{F_{L,y}}{F_{D,y}} \approx \frac{R^2 (\bar{u}_f - u_b)\left(\frac{\partial \bar{u}_f}{\partial y}\right)}{18 \upsilon (\bar{v}_f - v_b)}.$$

As this ratio goes to one, the bubble's y-position no longer changes as the forces are balanced, and as $\frac{\partial \bar{u}_f}{\partial y} = \frac{u_*}{\kappa y_b}$, there, $y_b \propto R^2$. Thus, Figure 47a is consistent and makes sense.

For the lower inset in the figure, gravity was discarded also, and the initial y_b position of the bubble was varied from 300 microns to 14 times the bubble radius of 150 microns with $u_b(t = 0) = 0$. Clearly the position of release strongly affects its final position in y.

In Figure 23 also from Sanders (2004), the effects of gravity are demonstrated. This is visible evidence that gravity is more important to the bubble trajectory for slower speeds as is intuitively understandable. In the figure, the hollow symbols are for zero gravity, the solid symbols represent the trajectory with gravity. Where gravity is present, buoyancy tended to return the bubbles to the surface. In fact when sufficient bubbles were present in the physical experiment under such a condition, the bubbles formed an air film. For the two higher speeds, buoyancy is relatively less

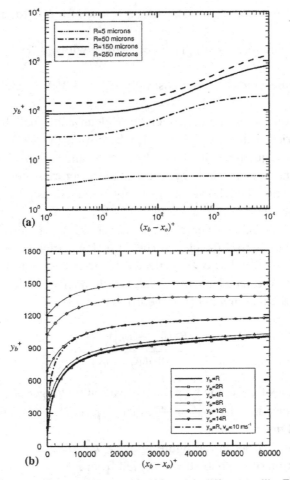

Figure 47. (a) Modeled trajectories of bubbles with different radii. Each bubble was released from rest at the surface 1.0 m downstream of the virtual origin of the boundary layer. The average bubble size from the experimental results is represented by the solid line ($R = 150$ microns). (b) The initial height ($y_{b,i} = 2R$ to $y_{b,i} = 14R$) and initial vertical velocity ($v_{b,i} = 0$ and $v_{b,i} = 10$ ms^{-1}) were varied, and the effects on the bubble trajectory are shown. The nominal condition (bubble released at $y_{b,i} = R$ where $R = 150$ microns for all cases, and $v_{b,i} = 0$ ms^{-1})) is represented by the solid line. In all cases shown in (a) and (b) $U = 18$ ms^{-1}, and gravity was neglected. In presenting these results, the both components of the trajectory are relative to the release position, and non-dimensionalized with the local wall unit, ν/u^*.

Figure 22. This figure that was extracted from Sander (2004) exhibits bubble trajectories determined using the simple bastardized model. Note that the two figures have different abscissa and ordinate scales with log for the upper and linear for the lower insets.

important and hence in the physical experiments the bubbles move from the plate with an attendant reduction in DR.

In general, buoyancy matters when its forces are the same order of magnitude as the vertical lift and drag components; at this point the equation ratios are approximately

$$\frac{F_{B,y}}{F_{D,y}} \approx \frac{R^2 g}{9 \upsilon (\bar{v}_f - v_b)}; \quad \frac{F_{B,y}}{F_{L,y}} \approx \frac{2g}{(\bar{u}_f - u_b)\left(\frac{\partial \bar{u}_f}{\partial y}\right)}.$$

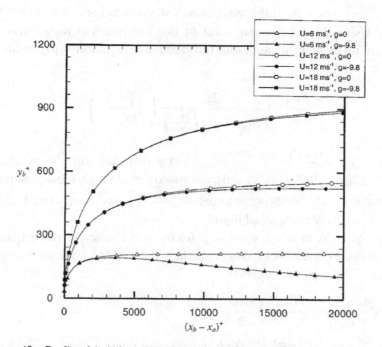

Figure 48. Predicted bubble trajectories with and without buoyancy effects, and with varying freestream velocities. All trajectories are for a bubble with $R = 150$ microns released from rest at the surface. Trajectories for each of the nominal flow speeds from the experiment are presented, with and without gravity. Where gravity is finite, buoyancy will tend to move the bubbles toward the surface. Trajectory components are non-dimensionalized as in Figure 47.

Figure 23. Bubble trajectories computed via the simple bubble model with and without buoyancy effects. For the lowest flow speed, gravity triumphs and the bubble returns toward the solid surface.

These equations imply that buoyancy forces dominate for lower speeds and larger bubbles. However, which one of the forces reigns supreme and when is not obvious (i.e. for a large bubble in a high speed flow or a small bubble in a slow flow). To get closer to an answer to this question, in a more recent manuscript (Elbing *et al.*, 2013), using these equations one can derive an equation of motion for the bubble in the y-direction with y the plate normal in the direction of gravity (This exercise was conducted to try and show how an air layer forms and to try to obtain scaling relationships.).

First, we note that the buoyancy force in the y-direction is: $F_{B,y} = \rho_w g \left(\frac{4\pi R^3}{3} \right)$ where ρ_w is the mass density of the water and R is the bubble radius. Using this relationship for $F_{B,y}$ one can return to their forces by determining them from their ratios (an alternative to returning to our earlier forms):

$$\frac{F_{B,y}}{F_{L,y}} \approx \frac{2g}{(\bar{u}_f - u_b)\left(\frac{\partial \bar{u}_f}{\partial y}\right)} \left(\frac{\frac{2\pi R^3}{3} \rho_w}{\frac{2\pi R^3}{3} \rho_w} \right)$$

and thus $F_{L,y} = \frac{2\pi R^3 \rho_w (\bar{u}_f - u_b)}{3} \frac{\partial \bar{u}_f}{\partial y}$. Likewise, from the ratio $\frac{F_{B,y}}{F_{D,y}}$, we obtain $F_{D,y} \approx 12\pi \rho_w \nu R(\bar{v}_f - v_b)$. Lastly, the added mass of the bubble (assuming potential flow) can be shown to equal $\frac{2\pi \rho_w R^3}{3}$ which is exactly one-half the mass of the bubble-displaced liquid.

Applying Newton's Second Law for the acceleration of the bubble in the y-direction, and remembering to include the bubble weight on the right-hand side of the equation yields

$$\left(\rho_A \frac{4\pi R^3}{3} + \rho_w \frac{4\pi R^3}{3} \right) \frac{dv_b}{dt}$$

$$= -\rho_w g \left(\frac{4\pi R^3}{3} \right) + \frac{2\pi R^3 \rho_w (\bar{u}_f - u_b)}{3} \frac{\partial \bar{u}_f}{\partial y} 12\pi \rho_w \nu R(\bar{v}_f - v_b)$$

$$+ \rho_A g \frac{4\pi R^3}{3}.$$

Here the subscript A represents air in the bubble and the subscript b represents the bubble. Dividing by $\left(\rho_A \frac{4\pi R^3}{3} \right)$ and noting that as demonstrated previously for inner variables scaling, $\frac{\partial \bar{u}_f}{\partial y} \approx \frac{\bar{u}_f}{y} = \frac{u_*}{l_v}$, the equation of motion upon using inner variables scaling becomes

$$\frac{\rho_w}{\rho_A}\left(\frac{\left(\bar{u}_f^+ - u_b^+\right)}{2} + \frac{9\left(\bar{v}_f^+ - v_b^+\right)}{R^{+2}}\right)$$

$$+ \left(1 - \frac{\rho_w}{\rho_A}\right)\left(\frac{vg}{u_*^3}\right) = \left(1 + \frac{\rho_w}{2\rho_A}\right)\left(\frac{dv_b^+}{dt^+}\right).$$

If we now use the fact that the mass density of air is 1/1000th that of water, multiply by ρ_A/ρ_w, cancel appropriately, then multiply by 2, we find after moving terms in v_b^+ to the left-hand side of the equation:

$$\frac{dv_b^+}{dt^+} + \frac{18}{R^{+2}}v_b^+ = -\frac{2vg}{u_*^3} + \left(\bar{u}_f^+ - u_b^+\right) + \frac{18}{R^{+2}}\bar{v}_f^+.$$

This equation is a first order, linear, constant coefficients ordinary differential equation that is inhomogeneous. A reasonable initial condition for the equation is $v_b^+(t=0) = 0$, which of course does not imply that the bubble is at $y^+ = 0$ nor does it mean that $u_b^+ = 0$. Using a standard solution technique (i.e. letting $v_b^+{}_{homog} = Ae^{\lambda t^+}$, finding a particular solution, and applying the initial condition to obtain the total solution) yields

$$v_b^+{}_{total} = (C)\left(1 - e^{-\frac{18}{R^{+2}}t^+}\right) \quad \text{with } C = -\frac{vgR^{+2}}{9u_*^3} + \frac{R^{+2}}{18}\left(\bar{u}_f^+ - u_b^+\right) + \bar{v}_f^+.$$

The bubble can rise ($v_b^+ < 0$ as per our coordinate system), fall, or remain in its initial, as yet unspecified position.

Assuming that our bubble will move toward the surface (i.e. rise), means that $v_b^+ < 0$, and therefore $C = -\frac{vgR^{+2}}{9u_*^3} + \frac{R^{+2}}{18}\left(\bar{u}_f^+ - u_b^+\right) + \bar{v}_f^+ < 0$ also. Moreover, Elbing *et al.* (2013) and our other LCC results had $R \sim 100l_v$, while $\bar{v}_f^+$ was scaled by u_*. Forming the ratio of these two terms as follows:

$$R^+ \sim 100 \text{ and } \bar{v}_f^+ = \frac{\bar{v}_f}{u_*} < 1 \text{ and hence } \frac{\bar{v}_f^+}{R^{+2}} \ll 1.$$

Therefore, neglecting the $\bar{v}_f^+$ term in C above, and then factoring R^{+2} from the equation, and recalling that our assumption was $v_b^+ < 0$, gives

$$\left(\bar{u}_f^+ - u_b^+\right) < \frac{2vg}{u_*^3},$$

a criterion for when the bubble moves toward the plate and independent of R^+ as long as $R^+ \sim 100$.

Although the left-hand side of this equation is not available *per se*, the terms should enable scaling of when an air layer forms through the bubbles moving toward the surface of the plate. That is, for results that obtain an air layer, $\frac{vg}{u_*^3}$ should scale the problem (The two has been dropped as unnecessary, and although in our experiments the v and g are constants, we retain them as variables.).

Using data from Elbing *et al.* (2008; 2013), the following equation was found:

$$\frac{q_{crit}}{q_s} = 6.135 \left(\frac{vg}{u_\tau^3} \right)^{-0.602}.$$

The data reduced as shown in the latter paper in their Figure 7 are reproduced below (Figure 24). It is clear that the available data collapse quite well.

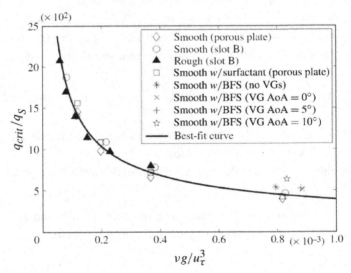

Figure 24. This figure, reproduced from Elbing *et al.* (2013), shows the reduced data and its collapse.

According to this analysis, any flux of gas that is on or exceeds this curve for a given value of the abscissa will result in an air layer. It must be remembered that the scaling resulted from a simple, bastardized bubble model that neglects many interactions, and simplifies the liquid flow as well.

Before returning to bubbles as a method of drag reduction and their transition to an air layer, we note that due to the lack of persistence downstream, it appears that BDR is not particularly feasible, except at slow speeds, unless as suggested by numerical analysis (e.g. Ferrante and Elghobashi, 2004) much smaller bubbles do in fact produce bubbles that remain in or persist in the inner layer farther downstream. There are the additional issues as to whether or not these much smaller bubbles can be produced readily in quantity for BDR, and whether they will even provide drag reduction.

To address these issues directly, the authors and co-workers conducted two sets of experiments in the mLCC at the University of Michigan's Marine Hydrodynamics Laboratory. In the first (Winkel *et al.*, 2004), freshwater, saltwater, and surfactant-laden water (Triton X100) solutions of flowing water were subjected to air injection in the TBL, and bubble-size distributions were quantified. Using high speed imagers to record frames of the bubbly flow with imager resolution determined using precise targets (pixel size ~12 microns), bubble sizes were determined manually (No suitable software with suitable accuracy was available to do this for our images that exhibited gray-scale variation for example.). This paper includes bubble histograms and lognormal fits to them, but here we present only their Table 2 which compared primarily mean bubble diameter and bubble diameter in viscous lengths for each liquid solution.

Table 1. Results from Winkel *et al.* (2004).

Table 2. Mean bubble diameters, mean bubble volumes, most probable diameters, and d^+ values

Solution	Mean bubble diameter (μm)	Mean bubble volume (mm^3)	Most probable diameter (μm)	$d^+ = d/l_\nu$
Baseline (tap water)	491 ± 18	0.109 ± 0.063	435 ± 10	323 ± 12
Surfactant (1 ppm)	420 ± 17	0.066 ± 0.024	385 ± 9	276 ± 11
Surfactant (5 ppm)	337 ± 7	0.032 ± 0.029	301 ± 4	221 ± 5
Surfactant (20 ppm)	215 ± 4	0.007 ± 0.003	205 ± 2	141 ± 3
Instant Ocean (9 ppt)	446 ± 12	0.063 ± 0.022	421 ± 6	293 ± 8
Instant Ocean (19 ppt)	194 ± 3	0.0047 ± 0.0013	187 ± 1	127 ± 2
Instant Ocean (33 ppt)	118 ± 3	0.0012 ± 0.0003	112 ± 1	78 ± 2
Instant Ocean (38 ppt)	113 ± 3	0.0011 ± 0.0005	105 ± 2	74 ± 2

As is evident from Table 1, a mean diameter range with a factor of about five was achieved. The concentrations of salt/Instant Ocean and Triton X100/surfactant are shown along with the baseline tap water.

In the second of these publications (Shen *et al.*, 2006) alluded to above, the influence of bubble size on BDR was investigated directly, also in the mLCC facility. Along with the three background solutions just mentioned, lipid-stabilized bubbles with mean diameters of 44 microns were injected. Wall-unit d^+ values spanned 200 to 18 in these experiments. The velocity profile measured agreed well with the 1/7th power law. Unfortunately, although an order of magnitude bubble-diameter difference existed between the tap water injection and those of the lipid-coated bubbles, the BDR was essentially the same for each as shown here.

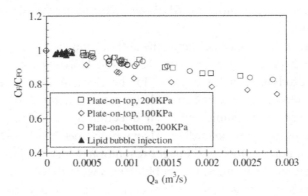

Fig. 14 Fraction of drag reduction with lipid bubble injection compared to nitrogen injection results in tap water

As is stated in Elbing *et al.* (2013), while the underlying physical mechanisms of BDR are not fully understood, it is generally agreed that bubbles reduce turbulent momentum exchange in the buffer region of the TBL. Most of the proposed drag reduction mechanisms involve a decrease in the near-wall Reynolds stress $\left(-\rho\langle u'v'\rangle\right)$, where ρ is the fluid density and u' and v' are the stream-wise and wall-normal velocity fluctuations, respectively. One mechanism for this reduction is a decrease in the bulk density, which would suggest that drag reduction would scale with the near-wall void fraction. This scaling is supported by the experimental results of Elbing *et al.* (2008). Another mechanism as suggested by Lumley (1973; 1977) is that the bubbles increase the local viscosity in the sub-layer and buffer region resulting in a reduction of the near-wall velocity fluctuations, which has been supported experimentally (Pal *et al.*, 1989) and computationally (Druzhinin and Elghobashi, 1998; Ferrante and

Elghobashi, 2004). Conversely, Nagaya *et al.* (2001) found experimentally that the turbulent fluctuations increased with gas injection. Hence an alternative mechanism was suggested that the bubbles decorrelate the stream-wise and wall-normal velocity fluctuations which results in a decrease in the Reynolds stress. These mechanisms as well as mechanisms related to bubble splitting and deformation (Meng and Uhlman, 1998; Lu *et al.*, 2005; van den Berg *et al.*, 2005) are supported in experiments as well as simulations, which suggests that multiple mechanisms are present. Thus it is not surprising that a universal scaling law has yet to be developed for BDR.

In spite of uncertainty about the physical mechanisms, the success of BDR at laboratory scale (models typically less than one-meter in length at downstream-distance-based Reynolds numbers at or below 10^7) motivated the extension of BDR measurements to larger scales. This is critical as the Reynolds number scaling of turbulent boundary layers is a nontrivial problem (Klewicki, 2010) that is further complicated due to the presence of multiple phases. One set of experiments used 12 to 50 m long models, slender flat-bottom ship models (Watanabe *et al.*, 1998; Kodama *et al.*, 1999; 2002) and one involved full-scale sea trials (Kodama *et al.*, 2000; Nagamatsu *et al.*, 2002; Kodama *et al.*, 2006). The most recent sea trials produced an overall power savings of a few percent. The first commercial production of bulk carriers that used an air injection drag reduction system (MALS, Mitsubishi Heavy Industries) was reported recently (Konrad, 2011). Three grain carriers are scheduled to be constructed with dimensions 237 m long, 40 m wide and 12.5 m draft (at design condition). Using available, limited information it appears that the air lubrication scheme will produce transitional air layers, which could reduce CO_2 emissions by 25% (as reported in news releases).

These ship-scale advances are encouraging, but due to the complexity of sea trials and minimal release of data from these commercial efforts, physical insights are not gleaned readily from such reports. For this we revert to the largest-possible laboratory experiments. The work of Sanders *et al.* (2006) and Elbing *et al.* (2008) produced two-dimensional experimental data at comparable model lengths to the current study within a laboratory setting, which enabled high fidelity, high Reynolds number BDR measurements. These results indicated that beyond 2 m downstream of the air-injection location, BDR produces minimal drag reduction as shear forces and turbulent motions remove bubbles from the near-wall region. Furthermore, injecting gas from multiple downstream locations produced either negligible change or a *decrease* in drag reduction relative to injection of the equivalent volume from a single location. As a consequence of the relatively poor downstream persistence of BDR, ALDR remains of interest despite the larger required air fluxes.

In the next chapter, we discuss the transition from BDR to ALDR as well as our work on air layer drag reduction.

References

Bogdevich, V.G. and Evseev, A.R. (1976). The distribution on skin friction in a turbulent boundary layer of water beyond the location of gas injection. Investigations of Boundary Layer Control (in Russian). *Thermophysics Institute Publishing House*, 62.

Bogdevich, V.G. and Malyuga, A.G. (1976). Effect of gas saturation on wall turbulence. Investigations of Boundary Layer Control (in Russian). *Thermophysics Institute Publishing House*, 49.

Ceccio, S.L. (2010). Friction drag reduction of external flows with bubble and gas injection. *Annu. Rev. Fluid Mech.* 42, 183–203.

Chang, E. and Maxey, M. (1995). Unsteady flow about a sphere at low to moderate Reynolds number. Part 2. Accelerated motion. *J. Fluid Mech.* 303, 133–153.

Clark III, H. and Deutsch, S. (1991). Microbubble skin friction reduction on an axisymmetric body under the influence of applied axial pressure gradients. *Phys. Fluids A* 3, 2948–2954.

Deutsch, S. and Castano, J. (1986). Microbubble skin friction reduction on an axisymmetric body. *Phys. Fluids* 29, 3590–3597.

Druzhinin, O.A. and Elghobashi, S. (1998). Direct numerical simulations of bubble-laden turbulent flows using two-fluid formulation. *Phys. Fluids* 10, 685–697.

Elbing, B.R., Winkel, E.S., Lay, K.A., Ceccio, S.L., Dowling, D.R. and Perlin, M. (2008). Bubble-induced skin-friction drag reduction and the abrupt transition to air-layer drag reduction. *J. Fluid Mech.* 612, 201–236.

Elbing, B.R., Makiharju, S., Wiggins, A., Perlin, M., Dowling, D.R. and Ceccio, S.L. (2013). On the scaling of air layer drag reduction. *J. Fluid Mech.* 717, 484–513.

Etter, R.J, Cutbirth, J.M., Ceccio, S.L., Dowling, D.R. and Perlin, M. (2005). High Reynolds number experimentation in the U.S. Navy's William B. Morgan Large Cavitation Channel. *Measurement Sci. Technol.* 16, 1701–1709.

Ferrante, A. and Elghobashi, S. (2004). On the physical mechanisms of drag reduction in a spatially developing turbulent boundary layer laden with microbubbles. *J. Fluid Mech.* 503, 345–355.

Fontaine, A.A. and Deutsch, S. (1992). The influence of the type of gas on the reduction of skin friction drag by microbubble injection. *Exp. Fluids* 13, 128–136.

Haberman, W.L. and Morton, R.K. (1953). An experimental investigation of the drag and shape of air bubbles rising in various liquids. David Taylor Model Basin.

Kawamura, T., Kakugawa, A., Kodama, Y., Moriguchi, Y. and Kato, H. (2003). Effect of Bubble Size on the Microbubble Drag Reduction of a Turbulent Boundary Layer. *Proc. ASME Fluids Engineering Division* Summer Meeting, 1–8.

Klewicki, J.C. (2010). Reynolds number dependence, scaling, and dynamics of turbulent boundary layers. *Trans. ASME: J. Fluids Eng.* 132(9), 094001.

Kodama, Y., Kakugawa, A. and Takahashi, T. (1999). Preliminary experiments on microbubbles for drag reduction using a long flat plate ship. ONR Workshop on Gas Based Surface Ship Drag Reduction, Newport, USA, 1–4.

Kodama, Y., Kakugawa, A., Takahashi, T. and Kawashima, H. (2000). Experimental study on microbubbles and their applicability to ships for skin friction reduction. *Int. J. Heat Fluid Flow* 21, 582–588.

Kodama, Y., Kakugawa, A., Takahashi, T., Nagaya, S. and Sugiyama, K. (2002). Microbubbles: Drag reduction mechanism and applicability to ships. *24th Symp. Naval Hydro.*, 1–19.

Kodama, Y., Hori, T., Kawashima, M.M. and Hinatsu, M. (2006). A full scale microbubble experiment using a cement carrier. *Eur. Drag Reduction and Flow Control Meeting*, Ischia, Italy, 1–2.

Konrad, J. (2011). The bubble ship — Mitsubishi's new green ship technology, gCaptain, Unofficial Networks, October 24, http://www.gcaptain.com.

Legendre, D. and Magnaudet, J. (1998). The lift force on a spherical bubble in a viscous linear shear flow. *J. Fluid Mech.* 368, 81–126.

Lu, J., Fernández, A. and Tryggvason, G. (2005). The effect of bubbles on the wall drag of a turbulent channel flow. *Phys. Fluids* 17, 095102.

Lumley, J.L. (1973). Drag reduction in turbulent flow by polymer additives. *J. Polymer Sci. Macromol. Rev.* 7, 263–290.

Lumley, J.L. (1977). Drag reduction in two phase and polymer flows. *Phys. Fluids* 20, S64–S70.

Madavan, N.K., Deutsch, S. and Merkle, C.L. (1984). Reduction of turbulent skin friction by microbubbles. *Phys. Fluids* 27, 356–363.

Madavan, N.K., Deutsch, S. and Merkle, C.L. (1985). Measurements of local skin friction in a microbubble-modified turbulent boundary layer. *J. Fluid Mech.* 156, 237–256.

Magnaudet, J. and Eames, I. (2000). The motion of high-Reynolds-number bubbles in inhomogeneous flows. *Annu. Rev. Fluid Mech.* 32, 659–708.

Maxey, M.R. and Riley, J.J. (1983). Equation of motion for a small rigid sphere in a nonuniform flow. *Phys. Fluids* 26, 883–889.

McCormick, M.E. and Bhattacharyya, R. (1973). Drag reduction on a submersible hull by electrolysis. *Naval Engrs J.* 85, 11–16.

Meng, J.C.S. and Uhlman, J.S. (1998). Microbubble formation and splitting in a turbulent boundary layer for turbulence reduction. *Intl Symp. on Seawater Drag Reduction*, 341–355.

Merkle, C.L. and Deutsch, S. (1992). Drag reduction in liquid boundary layers by gas injection. *Prog. Astronaut. Aeronaut.* 123, 351–412.

Migirenko, G.S. and Evseev, A.R. (1974). Turbulent boundary layer with gas saturation. *Problems of Thermophysics and Physical Hydrodynamics* (in Russian).

Nagamatsu, T., Kodama, T., Kakugawa, A., Takai, M., Murakami, K., Ishikawa., Kamiirisa, H., Ogiwara, S., Yoshida, Y., Suzuki, T., Toda, Y., Kato, H., Ikemoto, A., Yamatani, S., Imo, S. and Yamashita, K. (2002). A full-scale experiment on microbubbles for skin friction reduction using SEIUN MARU. Part 2: The full-scale experiment. *J. Soc. Naval Arch. Japan* 192, 15–28.

Nagaya, S., Kakugawa, A., Kodama, Y. and Hishida, K. (2001). PIV/LIF measurements on 2-D turbulent channel flow with microbubbles. *4th Int. Symp. on PIV*, Goettingen, Germany.

Pal, S. Merkle, C.L. and Deutsch, S. (1988). Bubble characteristics and trajectories in a microbubble boundary layer. *Phys. Fluids* 31(4), 744–751.

Pal, S., Deutsch, S. and Merkle, C.L. (1989). A comparison of shear stress fluctuation statistics between microbubble modified and polymer modified turbulent flow. *Phys. Fluids A* 1, 1360–1362.

Sanders, W.C. (2004). Bubble Drag Reduction in a Flat Plate Boundary Layer at High Reynolds Numbers and Large Scales. Doctoral Thesis, University of Michigan.

Sanders, W.C., Winkel, E.S., Dowling, D.R., Perlin, M. and Ceccio, S.L. (2006). Bubble friction drag reduction in a high-Reynolds-number flat-plate turbulent boundary layer. *J. Fluid Mech.* 552, 353–380.

Shen, X., Perlin, M. and Ceccio, S.L. (2006). Influence of bubble size on micro-bubble drag reduction. *Exp. Fluids* 41, 415–424.

Takahashi, T., Kakugawa, A., Nagaya, S., Yanagihara, T. and Kodama, Y. (2001). Mechanisms and scale effects of skin friction reduction by microbubbles. *Second Symp. on the Smart Control of Turbulence*, 1–9.

van den Berg, T.H., Luther, S., Lathrop, D.P. and Lohse, D. (2005). Drag reduction in bubbly Taylor–Couette turbulence. *Phys. Rev. Lett.* 94, 044501.

Watanabe, O., Masuko, A. and Shirose, Y. (1998). Measurements of drag reduction by microbubbles using very long ship models. *J. Soc. Naval Arch. Japan* 183, 53–63.

Winkel, E.S., Ceccio, S.L., Dowling, D.R. and Perlin, M. (2004). Bubble size distributions produced by wall-injection of air into flowing freshwater, saltwater, and surfactant solutions. *Exp. Fluids* 37, 802–810.

Chapter 4

Transition from BDR and Air Layer Drag Reduction

First, we discuss the transition from bubble drag reducing flows, where there are discrete small bubbles interacting with the underlying liquid turbulent boundary layer, to the formation of air layers, where bubbles have merged to form a high void fraction region between the outer flow and the solid surface. We then evaluate the implementation of air layer drag reduction (ALDR).

4.1. Transition from BDR to ALDR

Due to the poor persistence of bubbles in the inner layer of the boundary layer, and hence the rapid loss of drag reduction, BDR's prospects for implementation is quite limited; however fortuitously, it was found that with sufficient gas injection an air layer formed (ALDR) that was quite stable and afforded near 100% friction drag reduction on the HIPLATE test apparatus. The present discussion follows Elbing *et al.* (2008) closely.

The experiments were similar to others discussed previously, and thus we mention the salient features/differences only. Gas was injected at $X_{inj} =$ 1.38 and 3.78 m; two different types of injectors were tested with various configurations (the interested reader is referred to Elbing *et al.* (2008) for specifics); metering was done by mass-flow sensors as before; skin-friction balances measured drag on some tests while integral friction measurements were also made over six 1.6 m^2 plates mounted on rails located beneath the model skin and instrumented with load cells; bubble imagers were used (measured through a prism that extended through the plate and into the flow 5 mm); and lastly, electrical impedance (EI) probes were used to measure bulk void fraction of the near-wall flow. The data collected from these probes informs our discussion on the transition to air layer. Therefore, a brief discussion of EI probes is warranted.

The EI probes used were based on Cho *et al.* (2005). The diagram of each circuit and the equation governing it according to Kirchoff's law are as shown:

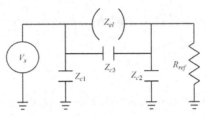

FIGURE 4. Circuit diagram of the electrical impedance probes used in Test 1. Shown is the voltage source, V_s, the impedance of the bubbly flow, Z_{el}, reference resistor, R_{ref}, and the stray capacitance from the lead wires Z_{c1}, Z_{c2} and Z_{c3}.

$$Z_{el} = \frac{(V_s - V_{ref})R_{ref}Z_{c2}Z_{c3}}{V_{ref}[Z_{c2}(R_{ref} + Z_{c3}) + R_{ref}Z_{c3}] - R_{ref}Z_{c2}V_s}. \qquad (2.1)$$

Here, the voltage source is denoted V_s, the impedance of the bubbly flow is Z_{el}, the reference resistor is represented by R_{ref}, and the capacitance from the lead wires by Z_{c1}, Z_{c2}, and Z_{c3}, respectively. To compute the void fraction, α, the following equation that represents the Maxwell mixture model (that assumes a uniformly disperse bubbly mixture — that did not exist) was utilized:

$$\alpha = \frac{(R_m - R_w)\left(2 + \frac{\sigma_g}{\sigma_w}\right)}{(2R_m + R_w)\left(1 - \frac{\sigma_g}{\sigma_w}\right)}.$$

In this equation, subscripts m represents the mixture, w represents the water, and g represents the gas. The impedance of the mixture, Z_{el} which is a complex variable for AC circuits, is measured by the first equation and is related to the average void fraction. In the latter equation, R_m is the real part of the impedance of the mixture, R_w that of the water. The conductivities, σ_g and σ_w are those of the gas, which is approximately zero, and that of the water, respectively. The complex-valued admittance of the mixture is assumed to have negligible imaginary part (i.e. the susceptance is approximately zero), so that the admittance equals approximately the conductance, σ. The conductivity dominates the equation when $\sigma_w \gg 0.44\,\mu\text{Scm}^{-1}$ (its usual value is $\sim 3.5\,\mu\text{Scm}^{-1}$). S is the unit of conductance, Siemens. With these probes, we can compare the measured local skin friction to the measured near-wall gas fraction for varying flow conditions.

As expected and can be seen in the three figures below (7–9) from Elbing *et al.* (2008) for 6.7, 13.3 and 20.0 ms^{-1}, large drag reduction occurs near injection for the BDR cases, but is lost rapidly downstream as the bubbles are pushed from the inner boundary layer. The key to the right of the plot

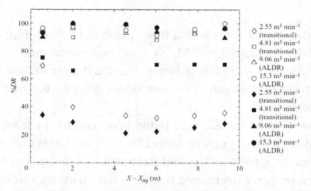

FIGURE 7. %*DR* versus $X-X_{inj}$ at a free-stream speed of 6.7 m s^{-1} during Test 1. A comparison of the Slot A (solid symbols) and porous-plate (open symbols) injectors at the four gas injection rates is presented. The key gives the volumetric gas injection rate corrected for test section static pressure, and in parentheses is whether this injection rate corresponds to BDR, ALDR, or the transition between BDR and ALDR.

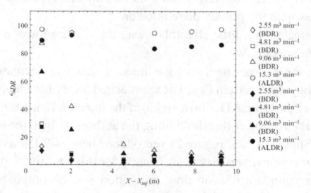

FIGURE 8. As figure 7, but for a free-stream speed of 13.3 m s^{-1}.

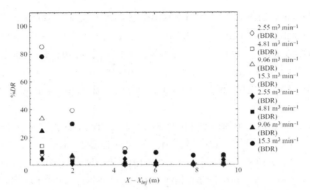

FIGURE 9. As figure 7, but for a free-stream speed of $20.0 \, \text{m s}^{-1}$.

in each of the figures notes which flows exhibited bubble drag reduction, transitional drag reduction, and ALDR as defined momentarily. It is clear in all cases of ALDR ($\sim 80\%$DR or larger) that the drag reduction is affected little by moving downstream. The transitional flows lie between these two extreme situations.

To show visually the adjacent liquid layer and the high void fraction adjacent layer cases, Figure 10 from Elbing *et al.* (2008) is reproduced. The photographs agree with the void fraction measurements by impedance probes when properly interpreted (i.e. both have dark regions beneath the plate).

For the $20 \, \text{ms}^{-1}$ free-stream speeds, presented in their Figure 12 reproduced here is the impedance-probe measured void fraction versus distance downstream from injection for the two injector types; the surfactant addition was with the porous-plate injector.

The surfactant has little effect; however the slot has lower α's than the porous plate injector.

During these experiments as gas injector flux was increased, it was found that beyond a certain flux, DR approached a very high level all along the surface of the HIPLATE downstream of the injector. Using 80%DR as an arbitrary but large lower threshold limit, the authors defined the occurrence of ALDR. Shown in their Figure 21 reproduced here, is %DR as a function of q for four downstream distances. Clearly for this $11.1 \, \text{ms}^{-1}$ free-stream speed, near complete friction drag mitigation was accomplished on the HIPLATE for a q-threshold value of approximately $0.066 \, \text{m}^2\text{s}^{-1}$.

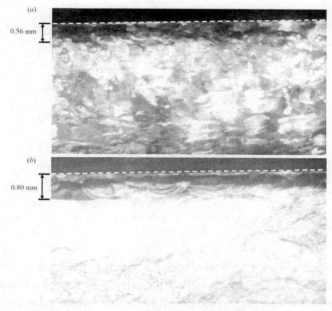

FIGURE 10. Photographic images of the near-wall bubbly flow at $13.3\,\mathrm{m\,s^{-1}}$ ($X = 5.94\,\mathrm{m}$) with (a) 5.1 and (b) $15.3\,\mathrm{m^3\,min^{-1}}$ injection from Slot A. The image is approximately 4.1 (wall-normal) by 7.7 (streamwise) mm, and the flow is from right to left. Both images have a dark region adjacent to the wall (region below the superposed white dashed line that represents the plate location) indicating that there are minimal air–water interfaces. (a) has a mean void fraction of 1 %, which indicates that the dark region is a liquid-layer. Conversely, (b) has large impedance measurements that prevented an accurate void fraction measurement, but the high impedance indicates that the dark region is primarily air.

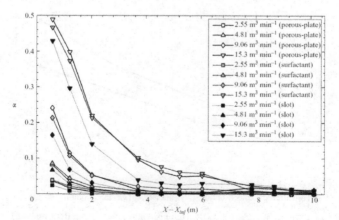

FIGURE 12. The near-wall void fraction, α, measured with electrical impedance probes versus downstream distance from the injector at the four nominal injection rates is shown for the porous-plate (with and without surfactant in the background) and upstream injection from Slot A without surfactant background. These data from Test 1 all correspond to free-stream speeds of $20.0\,\mathrm{m\,s^{-1}}$ and the large, 6.4 mm, electrode spacing.

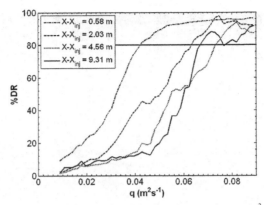

FIGURE 21: *%DR* versus *q*, the volumetric gas injection rate per unit span (m^2s^{-1}) from Test 1b for 11.1 ms^{-1} at four downstream locations. The solid horizontal line at *%DR* = 80% is the threshold used for defining ALDR.

In their Figure 23, for all flow speeds for which ALDR was achieved (as limited by the air compressor capacity), the critical air flux as a function of free-stream speed was presented, and it is shown below. Also presented is their Figure 26 that exhibits the threshold to ALDR curves for two experiments conducted roughly one year apart, and for which there is

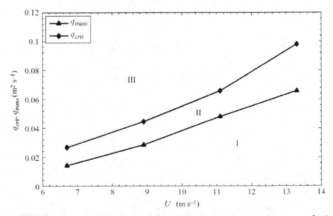

FIGURE 23. Transition gas injection rate, q_{trans} and critical gas injection rate, q_{crit}, (m^2s^{-1}) versus free-stream speed determined from Test 1b data. These two curves define the boundaries for the three drag reduction regions: I, BDR; II, transition region between BDR and ALDR; and III, ALDR.

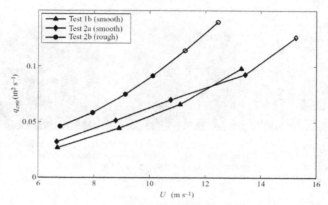

FIGURE 26. q_{crit} versus U_∞ for ALDR. Open symbols correspond to data points that exceeded the flow meter calibration range and are only estimates. Smooth-model ALDR experiments, Tests 1b and 2a were conducted about a year apart with different injector geometry. The roughened model, Test 2b, was tested immediately following Test 2a with the same injector.

agreement to within the error bars of the measurements. In addition, the latter figure also includes data recorded during their experiments on a rough surface, over which ALDR was achieved, but with a 30 to 50% increased flux penalty.

As definitive visual evidence that an air layer was formed, the authors presented Figure 27 that is once again reproduced herein. This set of three images shows the working surface of the roughened plate with free-stream speed $6.8\ \mathrm{ms}^{-1}$ during non-injection (left inset) in which the faint outline of an instrumentation hatch is visible and a clearly visible optical window may be seen. In the case of BDR, shown in the middle image, neither of these is visible; however in the right inset where an air–water interface exists beneath the plate surface, the optical window can again be seen. This is typical of visibility in these experiments: water with no interface and no air is quite transparent/clear; water with a gas void fraction of even 2 to 3% becomes nearly opaque in that region; while water with a wavy interface between air and water as occurs in ALDR is much more transparent than with BDR, but may exhibit distortion due to the light refraction that occurs through the wavy interface.

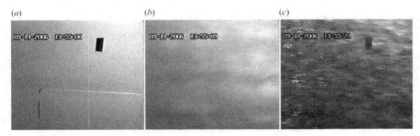

FIGURE 27. Images recorded during Test 2b of the roughened model surface at 6.8 m s^{-1}: (*a*) no injection; (*b*) bubble drag reduction; and (*c*) air-layer drag reduction. The streamwise (right to left in images) and spanwise (top to bottom in images) directions of the model were observed by the recording system.

4.2. Scaling of Air Layer Formation

Elbing *et al.* (2013) discussed the structure of air layers and the scaling of the critical gas flux necessary to achieve a layer. The data presented above show that the critical gas flux depends on the freestream speed as well as the level of surface roughness, with the required flux increasing with increases in each parameter. The transition from bubbly flow to a layer occurs when the buoyancy forces on individual bubbles are sufficient to move them toward the plate surface, overcoming shear induced lift forces that would move the bubbles from the surface. As presented in the previous chapter, Elbing *et al.* (2013) extended the scaling of Sanders *et al.* (2006) to examine the trajectory of a single bubble in the TBL shear flow. The equation of motion for the bubble integrated to show that the vertical velocity component of the bubble (e.g. toward the surface of the plate) was positive when

$$U_w^+ - U_b^+ < \frac{2vg}{u_\tau^3},$$

where u_τ and v are the friction velocity and kinematic viscosity of the single phase flow, U_w^+ and U_b^+ are the liquid and bubble flow velocities normalized with the friction velocity, here u_τ (e.g. inner scaled), and g is the gravitational acceleration. Hence, the bubbles will move from the plate when the buoyancy forces are overcome by the shear induced lift in the near-wall region. This scaling is used to re-plot the critical gas flux, and good collapse of the data was achieved.

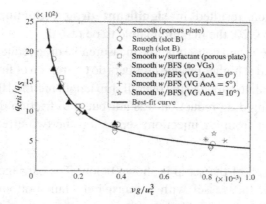

FIGURE 7. ALDR critical volumetric flux data from the current study and Elbing *et al.* (2008) scaled with the ratio of buoyancy to shear forces. Injection method, surface roughness and surface tension were varied.

We reiterate here that the best-fit curve for the scaled data was given by

$$\frac{q_{crit}}{q_s} = 6.135 \left(\frac{\upsilon g}{u_\tau^3}\right)^{-0.602}.$$

Note that we cannot predict the critical flux based on first principles, but we can scale it based on experimental observations.

4.3. Air Layers on Ship Models

In unpublished research, Gose and Perlin (2011) used an air layer beneath the Olympic Spirit model at the Marine Hydrodynamics Laboratory at the University of Michigan. Excellent results were obtained; however the results were never submitted for publication as the investigators were unable to parse the drag reduction from the propulsion (though not for lack of trying as six alternative techniques were attempted). The combined drag reduction and propulsive effects of the air injection exceeded the total drag, and the vessel was able to propel itself in the tow tank. Some of these results are presented.

Experiments were conducted at various model speeds for barehull and an one-inch strake and bow-step arrangement to determine total drag reduction due to air injection. The setup included a plate along the line of injection to direct the air downstream. The results are as follows:

(1) A 5% drag penalty was measured, from the added components, across all Froude numbers tested.

(2) Air injection resulted in **significant drag reduction** for Froude numbers to 0.20, the maximum able to be tested.

(3) Full air layers, and consequently, measureable drag reduction could not be measured at higher Froude numbers do to a region of flow separation. A new bow that extended the overall ship length rectified the separation.

(4) The resulting drag reduction was attributed to friction drag reduction and a thrust from air injection; separating the two effects was never achieved.

A photograph of the setup is shown below. The carriage, the blower–control system, the vessel with a transparent hull bottom for viewing purposes can all be seen in the figure.

Partial results from the model experiments are presented below. The upper inset shows three curves: the barehull total drag, the modified-for-air-injection barehull total drag (with about a 5% drag penalty as mentioned above), and the total drag with modified hull and air injection. As is evident, significant propulsion plus drag reduction are realized. The lower two insets, for Fr of 0.10 and 0.15, depict the regimes dominated by friction reduction and by thrust due to air injection. Unfortunately, the precise amounts due to each of these mechanisms were never distinguished.

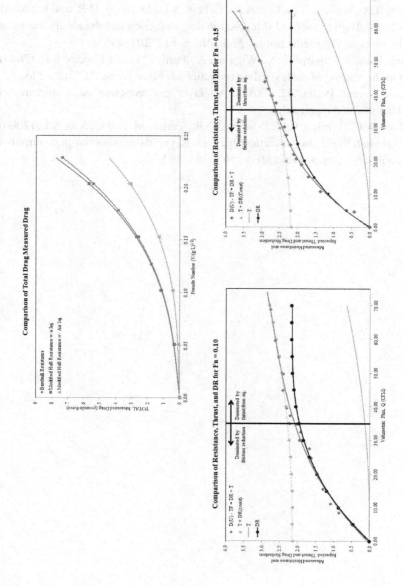

References

Cho, J., Perlin, M. and Ceccio, S.L. (2005). Measurement of near-wall stratified bubbly flows using electrical impedance. *Meas. Sci. Tech.* 16, 1021–1029.

Elbing, B.R., Winkel, E.S., Lay, K.A., Ceccio, S.L., Dowling, D.R. and Perlin, M. (2008). Bubble-induced skin-friction drag reduction and the abrupt transition to air-layer drag reduction. *J. Fluid Mech.* 612, 201–236.

Elbing, B.R., Mäkiharju, S.A., Wiggins, A., Perlin, M. and Ceccio, S.L. (2013). On the scaling of air layer drag reduction. *J. Fluid Mech.* 717, 484–513.

Gose, J.W. and Perlin, M. (2011). Air layer rag reduction on a ship model. Unpublished manuscript.

Sanders, W.C., Winkel, E.S., Dowling, D.R., Perlin, M. and Ceccio, S.L. (2006). Bubble friction drag reduction in a high-Reynolds-number flat-plate turbulent boundary layer. *J. Fluid Mech.* 552, 353–380.

Chapter 5

Friction Drag Reduction by Partial Air Cavities or PCDR

To begin our discussion of PCDR, Makiharju *et al.* (2010) (*Inter. Conf. Multiphase Flows*) is reviewed. The experimental setup included the HIPLATE, but with significant structural changes (over the one discussed so far) that facilitated the cavity (i.e. a backward-facing step (BFS) was added as well as a cavity-closing "beach"). In addition, the instrument suite was altered to include three laser Doppler velocimeters (LDVs) in addition to the upstream reference LDV, and three traverses that protruded into the flow and were perpendicular to the plate's working surface. The traverses included a set of sensor packages.

5.1. PCDR on the HIPLATE Model

The general experimental setup of the model, that was similar to Lay *et al.* (2010), and the free-surface-generating gate that could be fixed or articulated and created a downstream hydraulic jump are shown in the attendant figures (Figures 1 and 2, and Table 1 from Makiharju *et al.*, 2010). Also shown is the working surface with the instrument and measurement locations noted, and a table with the precise locations of the positions.

In addition to investigating the air fluxes required to establish and maintain the cavities at higher speeds than Lay *et al.* (2010), in Makiharju *et al.* (2010), closure geometry was altered to try to minimize air loss to the free-stream flow and to mitigate cavity unsteadiness. The effects of large flow perturbations via flap oscillation were studied also to quantify the robustness of the cavities. That is, how resilient the cavities were when subjected to forcing as they might experience in the open ocean such as

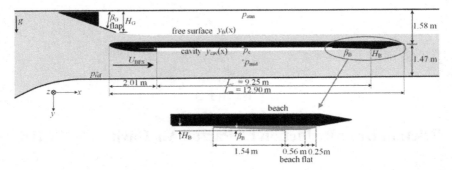

Figure 1. Profile of the gate and model in the LCC's test section. The origin of the coordinate system is at the base of the BFS, as detailed in Table 1. For clarity, the axes are shown shifted to the side and not at the true location of the origin. Inset: the cavity-terminating beach colored grey shown in detail. The upstream height of the cavity (i.e. the backward facing step) is 0.18 m. Notice that the flap is articulated and was oscillated during the experiments.

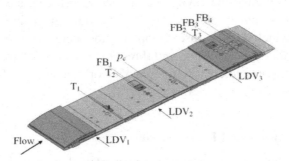

Figure 2. The model as seen from below with the measurement and instrument locations shown. The detailed locations are given in Table 1.

pressure fluctuations and large eddies (It is noted that the development and use of the gate afforded higher speeds as compared to Lay *et al.* (2010) as well as the use of surface pressure perturbations to study cavity response.).

The instrument suite included pressure transducers, electrical impedance (EI) probes, Pitot tubes, video imagers, LDVs, and force balances some of which were parallel to the flow, others of which were parallel to the beach surface. The specially-designed traversing mechanisms each had three probes: (1) Pitot-static probe, (2) time-of-flight electrode (similar physics to EI probe discussed previously but with five conducting

Table 1. Coordinates of the measurement and instrument locations. The base of the BFS is also the injector location. Results from all instrument locations are not presented, but all were used in interpretation of the results. On the beach the instruments are mounted flush with the local surface and as before $y(x)$ is the distance normal to the model surface.

Instrument/Object	x [m]	y [m]
Traverse 1	1.57	0.10...0.20
Traverse 2	3.86	0.10...0.20
Traverse 3	8.60	0.07...0.17*
Force balance #1	3.93	0.00
Force balance #2	8.58	0.07*
Force balance #3	9.04	0.08*
Force balance #4	9.56	0.09*
p_{ref} pressure tap	-2.43	1.18
p_{cav} pressure tap	4.60	0.00
p_{mid} pressure tap	4.70	0.26
LDV_{ref}	-2.04	0.27
LDV_1	-0.22	0.00...0.16
LDV_{BFS}	-0.22	0.04
LDV_2	3.93	-0.16...0.36
LDV_3	7.86	0.00...0.21
Leading edge	-2.01	-0.09
BFS/Injector	0.00	0.18/0.00

rings spaced at 5 mm on a 3.2 mm rod — first electrode via first, second, and third rings, while the second electrode used third, fourth, and fifth rings — speed of bubbles determined via cross correlation of two electrodes data), and (3) a point electrode probe. The entire mechanism could be traversed vertically from plate-flush to plate-proud, and it is shown below by the two Figures (3a and 3b) from Makiharju *et al.* (2010).

Two cameras recorded stills; high-speed video to capture the unsteady 3D closure, shapes, and oscillations were recorded by two imagers; and four synchronized imagers (speeds of 210 and 60 frames per second) recorded the cavity and flow from various stations positioned outside the test section. The pressure transducers monitored the ±5% variations created by the forced articulation of the gate when implemented.

The two primary quantitative findings from these experiments are shown below. In Makiharju *et al.* (2010), they were Figures 6 and 9. The first figure is similar to one presented later in this chapter, in the economics

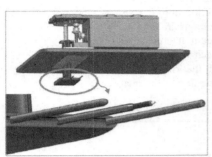

Figure 3a: Sketch of the traversing probes. All the probes had a 3.2 mm outer diameter and are shown in the enlarged view of the figure. The probes are from the left: time-of-flight electrode pairs, point electrode and Pitot-tube. The upper sketch shows the mechanism, enclosure and the plate that mounted into the model.

Figure 3b: Photo of traverse T_1 when the traverse was at y ~ 10 cm. The shiny areas on the surface are water droplets.

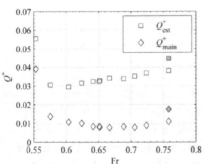

Figure 6: The minimum dimensionless gas flux Q^+ required to establish and maintain the cavity as a function of the Froude number. *Note: the velocity measurement, and hence the Froude number, for the last data points with grey fill, is in doubt because the LDV's data rate dropped drastically due to the presence of multiple small bubbles in the flow.*

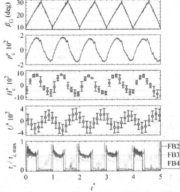

Figure 9: Time traces of the experiment with the oscillating gate where the cavity was not lost. The errorbars shown around the local mean values represent one standard deviation. From the bottom figure the cavity closure can clearly be seen to move upstream of FB4, then past FB3, and finally past FB2 prior to reversing a brief moment later. For the case shown: $\overline{H_c}$ ~ 17.0 cm, $\overline{p_c}$ ~ 6400 Pa, $\overline{U}$ = 5.46 m/s, Fr = 0.57, t_{cycle} = 15.7 seconds and Q^+~ 0.025. The non-dimensional values (° superscript) are defined in the text.

section, and exhibits the threshold of gas flux for the establishment and maintenance of the air cavities as a function of the Froude number.

Figure 9 of their publication depicts the effects due to gate-generated pressure fluctuations, i.e. pressure perturbations. The lowermost inset in this figure shows the cavity migration with time. Note that the cavity was not lost in the unsteady experiments when the gas flux was about twice that of the steady case.

For a qualitative description of the cavity closure, Figure 11 from that same paper is reproduced here. The six stages that resulted as a consequence of gate oscillation are shown; the phase lags are also evident.

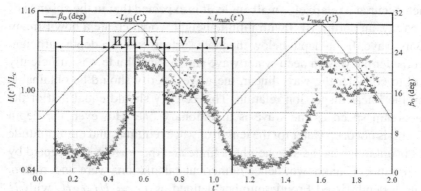

Figure 11: Cavity closure position oscillations and the low-pass filtered gate flap angle. Fr = 0.57. $L(t^*)$ notes the instantaneous closure position, which was visible within the camera's field of view. Subscripts min and max define the minimum and maximum closure position within the field of view. Subscript FB refers to the closure position at the spanwise coordinate of the force balances.

In figure 11, a presentation of the cavity closure position and the gate angle are both shown as a function of non-dimensional time, $t^* = (t - t_0)/t_{cycle}$. These curves also illustrate the phase lag that exists between them. The cavity position is derived from analysis of footage from the second 60 fps camera viewing the beach from beneath. The periodicity and shape of the closure's edge during the six stages of oscillations, that can be seen in figure 11, are described in table 2.

	Closure description	Shape
I	**Closure Near Beach Front** Upstream of the cavity closure (~20 cm), near the leading edge of the beach, the closure begins to exhibit violent churning and other three-dimensional disturbances. Ridges of wave patterns on the free surface are visible. The thickness of the closure edge is greater than for other stages. Air loss occurs through staggered and fairly regular cloud shedding. The spanwise correlation length of the clouds is fairly short (<10cm) and occurs at ~30Hz for the Fr = 0.57 case.	

II	**Uniform Cavity Growth** The cavity grows with spanwise uniformity. The trailing edge of the cavity has thinned and assumed a glassy scalloped appearance. The volume of air loss has noticeably decreased. The spanwise correlation length of the clouds being shed has increased.	
III	**Rapid Cavity Growth** As the cavity approaches the beach flat, the edge continues to thin and "fingers" and other spanwise non-uniformities appear. The fingers are spaced ~ 16 cm apart and extend to 50 cm in length. This stage is short lived.	
IV	**Cavity Extent Reaches Maximum** Corresponding with maximum gate angle, the cavity now extends almost the full length of the beach flat. The cavity edge shape fluctuates rapidly. There is erratic cloud shedding with long spanwise correlation.	
V	**Cavity Begins to Retreat** Necking of the fully extended closure region initiates near the sidewall. This results in substantial variations in closure position.	
VI	**Rapid Cavity Retreat** The extended part of the closure region detaches and divides. The cavity edge retreats to the position of Stage I with relative spanwise uniformity.	

Table 2: Typical cavity shapes and behavior during the six stages of oscillations shown in figure 11.

Determining the particular shape of the cavity is not trivial, but guidance exists, specifically from the work conducted in the former Soviet Union in the 1970s and 1980s. As expected, the depth, length, and beach angle are all fundamental to the resulting cavity shape.

In the case of the experiments by Makiharju *et al.* (2010), the cavity design objective was to lengthen it as much as possible within the constraints of the Large Cavitation Channel facility, and to have the closure on the beach occur at a relatively small angle. It is expected that in the absence of the cavity, an abrupt BFS with infinite height would create a downstream gravity wave. If the depth below the plate surface was at a depth sufficient to consider the wave a deep-water wave and the amplitude was sufficiently small to consider the wave linear, the wave celerity should be obtainable from the linear dispersion relation. This celerity should equal that of the free-stream speed as the wave is a stationary one. However, the depth (1.23 m) is not deep water for the length of wave required, and the amplitude (that is unknown as the step height is finite and the wave is intercepted by the structure) is unknown; hence numerical calculations were undertaken. Using a depth-based Froude number defined as $Fr = U/\sqrt{gH}$, with H the water depth, implied that the flow would have been supercritical with $1.4 \leq Fr \leq 2.7$, and thus indicated an undular to weak hydraulic jump region and even an oscillating jump for the largest Fr.

Moving to scaling of the partial cavity and its subsequent drag reduction, a list of dimensionless parameters are presented first see their Table 3.1. This set of parameters is required for our discussion and is gleaned from Makiharju (2012). Note that 11 dimensionless quantities are required in general.

To determine how to scale the problem, obviously multiple experiments at various scales are necessary; hence another set of tests was undertaken at the University of Michigan in the mLCC. Recall that this facility is a $1/14^{\text{th}}$ geosim model of the LCC (except that a 1.2 m extension had been added to each vertical leg to postpone cavitation to increased impeller rotation rates).

In these sets of experiments, the important relevant variables are shown in the following Table 3.2, while dimensionless groups that were the same are the ratios of densities and viscosities, non-dimensionalized step height, and ratio of the span of the cavity to its longitudinal extent.

Table 3.1 – Dimensionless groups.

Group	Definition	Note
Dimensionless gas flux	$q = \dfrac{Q}{UWH_{step}}$	Gas flux divided by free stream speed, model span and step height to non-dimensionalize
Reynolds number	$Re = \dfrac{\rho_l U L_c}{\mu_l}$	Ratio of inertia to viscous forces
Froude number	$Fr = \dfrac{U}{\sqrt{g L_c}}$	Ratio of inertia to gravity forces
Weber number	$We = \dfrac{\rho U^2 H_{step}}{\sigma}$	Ratio of inertia to surface tension forces
Density ratio	$\dfrac{\rho_g}{\rho_l}$	For the small scale experiments $\sim 1/830$
Viscosity ratio	$\dfrac{\mu_g}{\mu_l}$	For both large and small scale experiments ~ 0.018
Non-dimensionalized step height	$\dfrac{H_{step}}{L_c}$	For both large and small scale experiments ~ 0.019
Non-dimensionalized cavity span	$\dfrac{W}{L_c}$	For both large and small scale experiments ~ 0.33
Angle of closure surface with respect to free stream direction at infinity	β_B	Fixed at $1.7°$ for the small scale experiments
Fluctuation intensity	$\dfrac{u'}{U}$	
Non-dimensionalized length of disturbances	$\dfrac{\lambda}{L_c}$	Could be more meaningful to substitute L_c with closure wake thickness, t_c.

Table 3.2 – Parameter ranges of the experiments

	Large Scale	Small Scale
Fr	$0.5 - 0.9$	$0.5 - 1.0$
We	$(0.6 - 1.4) \times 10^6$	$(0.3 - 2.2) \times 10^3$
Re	$(4.6 - 6.9) \times 10^7$	$(0.9 - 1.6) \times 10^6$
U [m/s]	$5 - 7.5$	$1.3 - 2.5$
Q [m³/s]	$(0.3 - 2.6) \times 10^{-1}$	$(0.8 - 126) \times 10^{-6}$
L_c [m]	9.25	0.66

Under these assumptions, dimensional analysis suggests the following:

$$\frac{Q}{UWH_{step}} = f\left(Re,\ Fr,\ We,\ \frac{u'}{U},\ \frac{\lambda}{L_c}\right)$$

where the parameters and variables have been defined in Tables 3.1 and 3.2. Presumably, $\frac{u'}{U}$ and $\frac{\lambda}{L_c}$ are also functions of (Re, Fr, We), and so the following scaling relationship is used: $\frac{Q}{UWH_{step}} = f(Re, Fr, We)$. Again the ranges for the three independent non-dimensional numbers are shown above.

As in the larger model experiments, the smaller model experiments had an upper free-surface-forming gate upstream, and a second lower free surface beneath the plate due to the presence of the cavity. Figure 3.3 showing the plate with a partial cavity is shown here as extracted from Makiharju *et al.* (2012).

Figure 3.3 – Photo of the model and gate in the test section of the MLCC. In addition to the free surface of the cavity behind the BFS, we also have a second free surface above the plate and originating from the leading edge of the gate. The second free surface terminated in the tunnel's diffuser *via* a hydraulic jump.

Conducting the experiments without and with a roughness boundary layer trip did not alter the required flux, Q noticeably in the mLCC. Therefore, for at least small turbulent boundary layer changes, the cavity was somewhat insensitive. A primary difference between the large scale and the small scale experiments was that in the latter case, with slow free-stream speeds, a completely closed cavity could be achieved as in Arndt *et al.* (2009).

Quantitatively, Makiharju observed that the establishment flux in both the LCC and the mLCC was much larger than the maintenance flux, and that the Froude number is clearly not sufficient to scale the fluxes. Both of these observations are abundantly clear in his Figure 3.11 reproduced on the next page.

To examine the surface tension effects on the closure in the small-scale experiment, which was not done in the LCC, Triton X100 surfactant was used so that the Weber number spanned 300–2200. As surfactant

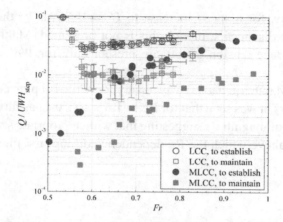

Figure 3.11 – Non-dimensionalized critical gas flux as a function of Froude number at different size scales. For all MLCC establishment fluxes at the lower Fr, which do not have a data point for the maintenance flux, this was below the range of the flow meters ($<10^{-3.4}$), but larger than zero.

is added, surface tension decreases. Recall that treated water at 20°C is quoted usually to have a surface tension of $0.072\,\mathrm{Nm}^{-1}$. The effect of the decrease in surface tension due to the surfactant is to increase the required gas flux. Here we show those results for a We change by a factor of ~ 2 with $U = 1.45\,\mathrm{ms}^{-1}$.

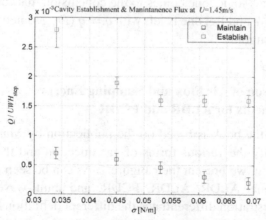

Figure 3.15 – Required cavity maintenance and establishment gas flux as a function of surface tension at $U = 1.45$ m/s.

Interestingly, for a speed increase to $U = 1.75\,\mathrm{ms^{-1}}$, the apparent *We* dependence disappears, though this is not presented. Makiharju concluded that surface tension effects are present only for $We < 600$ and $Fr < 0.7$.

Although Makiharju had no way to vary Re while keeping constant the other parameters, it was seen that for $Re > 1.5 \times 10^6$, the gas fluxes were of the same order of magnitude suggesting that with appropriate scaling there may be a flux above which the Re-dependence disappears. Figure 3.18 is presented here.

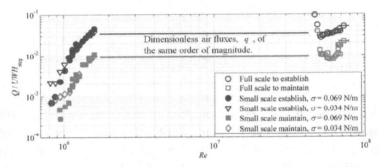

Figure 3.18 – Critical dimensionless gas fluxes as a function of Reynolds number.

Following this logic, one comes to the conclusion that for sufficiently large values of *We* and Re, one is left with $q = q\,(Fr,\text{geometry})$. Efforts in this regard are ongoing.

5.2. Comparison of Air Flux and Pumping Energy Requirements for ALDR and PCDR

This section of the book is based on the publication by Makiharju *et al.* (2012). To clarify the various forms of gas injection and the subsequent physical situation, we present their Figure 1. As can be seen in the figure, BDR, Transitional ALDR, ALDR, PCDR, and multi-wave PCDR are shown. This illustration differentiates the different definitions clearly.

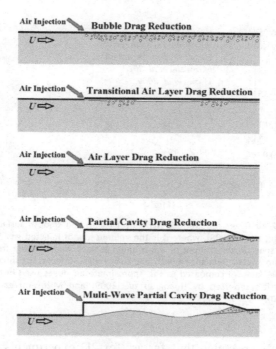

Fig. 1 Conceptual sketches illustrating the different air lubrication techniques. From the top: BDR, transitional, ALDR, PCDR and multi-wave PCDR.

From air layer drag reduction studies, we have the critical air fluxes (q in m²s⁻¹) as a function of the speed of the free stream. Hoang *et al.* (2009) and Mizokami *et al.* (2010), both from sea trials data, have been added to the data from Elbing *et al.* (2008) and are shown in Figure 2 on the next page.

In assessing the data presented in this figure, it should be realized that the flux data and the sea conditions during the trials of Hoang *et al.* and Mizokami *et al.* were not available. In fact in the Hoang *et al.* data the ship's trim and motions, flow around the hull, etc. may in fact have contributed to the existence of a transitional rather than complete air layer. Additionally as seen in Lay *et al.* (2010) upstream perturbations can cause the need for larger gas fluxes, and these are likely to be present on actual ship hulls. Lastly,

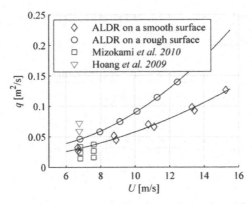

Fig. 2 Air fluxes required for ALDR over rough and smooth surfaces (Elbing et *al.*, 2008), where *q* is the volume flux of injected gas per unit span on the bottom of a horizontal surface, and *U* is the free-stream speed. Data for smooth and fully rough surfaces are shown. These data are compared to the approximate air fluxes used in the sea-trials reported by Hoang et *al.* (2009) and Mizokami et *al.* (2010).

the experiments presented thus far are for 2D experiments, where ships and the flows around them are necessarily 3D. Regardless in both of the sea trials drag reduction was realized at respectable levels. Hoang *et al.* realized 11% and 6% overall drag reduction for the ballast and full-load conditions, respectively (i.e. which they reported yielded 7% and 4% net **energy** savings) while Mizokami *et al.* reported **energy** savings of 8% to 12%.

The next figure exhibits the volumetric flow rate per unit span as a function of U, the free stream speed. The darker symbols and error bars represent the required values to establish the ventilated partial cavity (see Makiharju *et al.* (2012), Figure 1), the lighter represents the values required to maintain the established cavity. These data were recorded from experiments also in the LCC on the HIPLATE. They are reprinted from Makiharju *et al.* (2010). The shapes of the data curves in this figure are due to the following: On the left side, the gas flux increases dramatically due to the likely closure of the cavity upstream on the beach (i.e. distance upstream from where it was designed to close); on the right side, the gas flux increases due to ever-increasing gas flow downstream from the cavity closure region on the beach.

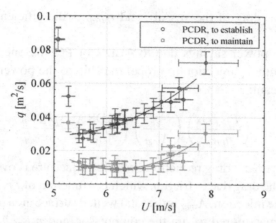

Fig. 3 The minimum gas volume flux per unit span, *q*, required to establish and maintain a ventilated partial cavity as a function of the flow speed (Mäkiharju *et al.*, 2010).

Partial cavities differ from ALDR layers in that there is a larger flux required to create the cavity (to displace the water that occupies the cavity) than to maintain the cavity. Additionally, the reason the fluxes increase for smaller *Fr* is that the cavity closes at larger angles which leads to excessive air entrainment loss from the cavity. PCDR's economic analysis along with that of ALDR is discussed next.

Now, the energy cost–benefit analysis is discussed for ALDR and PCDR. Obviously, (1) the air only reduces friction on the portion of the surface with which it is in contact, and this will be assumed. We also assume that (2) the projected area of the ship/vessel is not altered by the injected flow, and (3) that the ancillary accoutrements (strakes, small step for the air inlet, etc.) do not alter the drag on the vessel.

To evaluate savings via drag reduction, we consider only the portion of the ship's power used for propulsion and compare it to the power required to inject the air. Thus we have

$$\%E_{saved} = \left(\frac{\Delta t P_{saved}}{\Delta t (P_D/\eta_{prop})}\right)(100)$$

where $\%E_{saved}$ is the energy savings according to our assumptions, P_{saved} is the net power saved, P_D is the power required to propel the ship (i.e. to overcome the total drag of the vessel), and η_{prop} is the efficiency of the propeller.

To obtain the power savings due to ALDR or PCDR, one takes the power required with air injection to propel the ship to the power required to inject the lubricant:

$$P_{saved} = \left(\frac{P_D f_{FD}}{\eta_{prop}} \right) \left(\frac{A_{ac}}{A_{wet}} \right) \left(\frac{\%DR}{100} \right) - \frac{P_{comp}}{\eta_{elect}}.$$

Here %DR is the percent drag reduction over the surface area covered, f_{FD} is the fraction of the drag due to friction which is a function of Fr, A_{ac} is the area covered by the injection, A_{wet} is the total wetted surface area of the hull, P_{comp} is the power required to run the compressor, and η_{elect} is the ratio of electric power production efficiency (to run the blowers/compressors) divided by the propeller shaft power (If the propeller is electric powered, then $\eta_{elect} \rightarrow 1$.).

If we assume that the ship's Froude number ($Fr \equiv U/\sqrt{gL}$ with U, the ship speed and L, the overall length) is approximately 0.2, the drag due to friction is $\sim 60\%$. In the numerator of the last equation, the term $P_D f_{FD}$ represented P_{FD}, the power required to overcome the frictional drag. Likewise, as is usually done for ship propulsive calculations, the power required to overcome friction is calculated as that required for a flat plate with equal wetted surface area to that of the ship or $P_{FD} = \left(\frac{1}{2}\rho U^2 \right) (C_D WL) U$ where C_D is the drag coefficient, $W \times L$ gives the appropriate wetted area, and the first (dynamic pressure) and the last (ship speed) terms complete the expression.

Next, P_{comp}, the power to run the compressor/blower is needed. Assuming that the air compression is a polytropic process (i.e. pressure $\times$ (volume)^n = constant),

$$P_{comp} = \frac{\dot{m}_g p_1 n}{\eta_c \rho_{g,1}(n-1)} \left\{ \left(\frac{p_2}{p_1} \right)^{(n-1)/n} - 1 \right\}.$$

Here $\dot{m}_g$ is the mass flow rate of the gas (i.e. dm_g/dt), η_c is the compressor efficiency, p_1 and p_2 are the initial and final pressures of the gas, here the latter is a function of the draft of the vessel and the mass density of the liquid (salinity, temperature also), $\rho_{g,1}$ is the initial density of the gas, and

n can be replaced by $k = 1.40$ for air under the further assumption that the process is isentropic (n = polytropic index, k = ratio of specific heats).

To determine $\dot{m}_g$ (it is assumed that the gas is cooled to 25°C following compression), we may use $\dot{m}_g = Q\frac{\rho_{g,1}p_3}{p_1}$ where p_3 is assumed the hydrostatic pressure ($=\rho g D$ with D the draft of the vessel and the mass density of the liquid sufficient for this calculation). Additionally $p_2 = p_3 + p_{loss}$ with p_{loss} the pumping losses. Q is the volumetric flow rate of air/gas as obtained from q-values published in Elbing *et al.* (2008) and discussed previously or from Figure 2 just presented. Here, Q is determined as follows.

For ALDR, to estimate Q using Figure 2, $q = Q/W$, and quadratic curve fits with $R^2 = 0.99$ and $R^2 = 1.00$ on smooth and rough surfaces, respectively, equations are

$$\frac{Q}{W} = (5.01 \times 10^{-4})U^2 - (2.98 \times 10^{-5})U + 0.008$$

$$(\text{Smooth}; 6.7 \leq U \leq 15.3 \, \text{ms}^{-1})$$

and

$$\frac{Q}{W} = (1.26 \times 10^{-3})U^2 - (7.55 \times 10^{-3})U + 0.039$$

$$(\text{Rough}; 6.8 \leq U \leq 12.5 \, \text{ms}^{-1}).$$

For PCDR (to be discussed completely in a subsequent chapter) establishment flux and maintenance flux with $R^2 = 0.95$ and $R^2 = 0.81$, respectively, are given from curve fits as

$$\frac{Q}{W} = (4.76 \times 10^{-3})U^2 - (0.04796)U + 0.150$$

$$(\text{Establishment}; 5.5 \leq U \leq 7.5 \, \text{ms}^{-1})$$

and

$$\frac{Q}{W} = (7.01 \times 10^{-3})U^2 - (0.0866)U + 0.277$$

$$(\text{Maintenance}; 5.5 \leq U \leq 7.5 \, \text{ms}^{-1}).$$

Using these expressions, the $\%E_{saved}$ can be determined once a particular ship is considered. As an example, the American Spirit that operates on the Great Lakes is chosen. This ship has an overall length, L of 360 m, a beam,

W of 32 m, and a draft, D of 8.8 m. The assumptions that we have made that are all conservative include: % wetted area with air lubrication of 50%, a Fr of 0.2 that gives a f_{FD} of 0.60, %DR of 80%, η_{prop} of 75%, η_{elect} of 90%, and η_c of 60%. Next (the biggest assumption that must be made for ALDR), it is assumed that the air layer extends along the entire length of the hull, and that for PCDR, a multi-wave cavity requires equivalent air flux as a single wave cavity. Using these assumptions, Makiharju *et al.* (2012) Figure 4 was produced, and it is presented here. Note that the ordinate range is 13–22%.

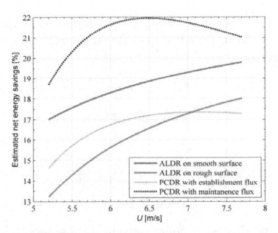

Fig. 4 Estimates of the potential net energy savings for a ship similar to the M/V American Spirit.

Reassuringly, considering all of the assumptions made, some gross in fact, these savings agree with the numbers reported by Hoang *et al.* (2009) from the *Pacific Seagull* (5–10%DR via ALDR or close to it) and by Mitsubishi Heavy Industries, 8 to 12%. Likewise for PCDR, 15% reported by Marin, and 20 to 25% by Stena.

References

Arndt, R.E.A., Hambleton, W.T., Kawakami, E. and Amromin E.L. (2009). Creation and maintenance of cavities under horizontal surfaces in steady and gust flows. *J. Fluids Eng.* 131, 111301.1–111301.10.

Elbing, B.R., Winkel, E.S., Lay, K.A., Ceccio, S.L., Dowling, D.R. and Perlin, M. (2008). Bubble-induced skin-friction drag reduction and the abrupt transition to air-layer drag reduction. *J. Fluid Mech.* 612, 201–236.

Hoang, C.L., Toda, Y. and Sanada, Y. (2009). Full Scale Experiment for Frictional Resistance Reduction using Air Lubrication Method. *Proc. of the Nineteenth International Offshore and Polar Engineering Conference*, June 21–26, pp. 812–817. Osaka, Japan.

Lay, K.A., Yakushiji, R., Makiharju, S., Perlin, M. and Ceccio, S.L. (2010). Partial cavity drag reduction at high Reynolds numbers. *J. Ship Res.* 54(2), 109–119.

Makiharju, S., Elbing, B.R., Wiggins, A., Dowling, D.R., Perlin, M. and Ceccio, S.L. (2010). Ventilated partial cavity flows at high Reynolds numbers. *Int. Conf. on Multi-phase Flows*, Tampa, FL.

Makiharju S. (2012). The Dynamics of Ventilated Partial Cavities over a Wide Range of Reynolds Numbers and Quantitative 2D X-ray Densitometry for Multiphase Flow. Ph.D. Dissertation, University of Michigan, Ann Arbor, Michigan, USA.

Makiharju, S., Elbing, B.R., Wiggins, A.D., Schinasi, S., Vanden-Broeck, J.-M., Perlin, M., Dowling, D.R. and Ceccio, S.L. (2013). On the scaling of air entrainment from a ventilated partial cavity. *J. Fluid Mech.* 732, 47–76.

Makiharju, S.A., Perlin, M. and Ceccio, S.L. (2012). On the energy economics of air lubrication drag reduction. *Inter. Jour. Nav. Arch. Ocean Engn.* 4(4), 412–422.

Mizokami, S., Kawakita, C., Kodan, Y., Takano, S., Higasa, S. and Shigenaga, R. (2010). Experimental study of air lubrication method and verification of effects on actual hull by means of sea trial. *Mitsubishi Heavy Industries Technical Review* 47(3), 41–47.

Part II

Passive Techniques

Chapter 6

Drag Reduction by Super and Partial Cavities

This chapter is concerned primarily with super and partial cavities on axisymmetric bodies. In addition, as discussed previously, we present material on two-dimensional partial cavities on bodies with nominally horizontal surfaces. As one of the authors (SC) recently published a paper on this subject, we follow closely that presentation (Ceccio, 2010). Axisymmetric cavity flows with gas injection are not discussed in detail herein; as excellent reviews are available (Knapp *et al.*, 1970; Wu, 1972; May, 1975; Brennan, 1995) we discuss them only briefly.

To begin, we present Figure 1 of Ceccio (2010). Here a disk is the "cavitator" with diameter d. The body has a total length, L, and the cavitator and body are in a freestream flow with speed, U, and ambient pressure, P_O. As can be seen in the figure, as the flow moves about the body, it separates downstream and this wake can be filled with gas to form a cavity. The pressure in the cavity is taken as P_C although the flow within the cavity causes pressure deviations from P_C. Of course the lowest pressure that the cavity can have is the liquid vapor pressure, P_V (Note that if the cavity is caused by pressure decrease due to vapor pressure, the cavity is termed a "natural" cavity.). As is shown pictorially in Figure 1, a cavity can envelop part (partial cavity) of the body, or it can extend downstream of the entire body for some distance (this cavity is termed a "super" cavity). In the first case, $L_C < L$ while in the latter instance, $L_C \gg L$.

We proceed by defining the usual cavitation number as

$$\sigma = \frac{P_O - P_C}{1/2\rho U^2}$$

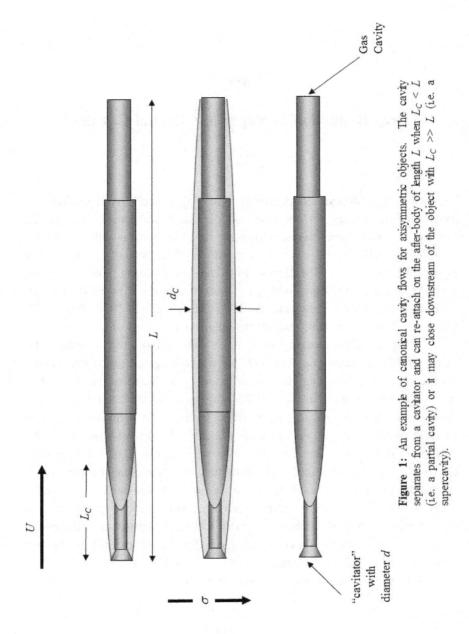

Figure 1: An example of canonical cavity flows for axisymmetric objects. The cavity separates from a cavitator and can re-attach on the after-body of length L when $L_C < L$ (i.e. a partial cavity) or it may close downstream of the object with $L_C >> L$ (i.e. a supercavity).

and a vapor cavitation number as

$$\sigma_V = \frac{P_O - P_V}{1/2\rho U^2}.$$

Recall that the cavitation number is similar to a pressure coefficient and if multiplied by an area/area is a ratio of pressure force/inertia force. If $\sigma < 1$, attention is required as cavitation may occur in regions of low pressure. In addition as buoyancy is likely important, we may scale with a diameter-based Fr (Froude number) where the length scale is chosen as the cavitator diameter, d, or as the body diameter. Choosing the former we have $Fr = U/\sqrt{gd}$, and assuming that gas is injected, its non-dimensional flux is

$$C_Q = \frac{Q}{Ud^2}$$

where as usual, Q is the volumetric flux with dimensions of length^3/time. Using the projected area of the body's cavitator, $\pi d^2/4$, the drag coefficient assumes the form

$$C_D = \frac{F_D}{\left(\frac{\pi d^2}{4}\right)\frac{1}{2}\rho U^2}$$

with F_D the drag.

According to Brennan (1995), the notion of an ideal (axisymmetric) cavity can be traced to the work of Kirchhoff (1869), although he did not mention cavities, Kirchhoff was investigating free streamlines of wakes. There are naturally occurring cavities of water vapor as mentioned above, and in addition there are so-called ventilated cavities where a non-condensable gas (air) is either injected or is included due to the liquid flow entraining air from the atmosphere (e.g. a high-speed planing boat "V"-hull as it re-enters the free surface and draws a cavity with it).

In Figure 2 taken from Schauer (2003), we see several photographs of a cavitator and its subsequent cavity as a function of decreasing σ. As $\sigma \to 0$, the cavities become more violent and developed, and so the non-dimensional flux, C_Q, increases. As the cavity forms initially, $\sigma \approx 0.24$, and a bubbly mixture forms (as C_Q is insufficient to create a continuous cavity). However, as σ is decreased a continuous cavity is evident (Decreased σ can be accomplished by lowering the ambient pressure via drawing a vacuum,

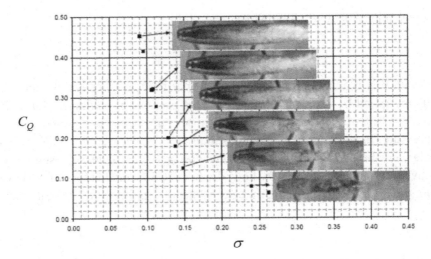

Figure 2: A series of images showing of a cavity forming on a 1cm disk cavitator for decreasing cavitation numbers, σ, with the corresponding value of flow coefficient, C_Q (Schauer 2003).

all at constant freestream speed, U, or by increasing U while maintaining a constant P_O.).

The cavity shape is of course of interest, as is the drag coefficient. Both are a function of the dimensionless variables, σ, Fr, and C_Q, that were given previously. For the moment, let us assume that the gas flux is constant and, as the cavity increases with decreasing σ, it is intuitive that the drag coefficient increases with increasing σ (as the cavity wanes!).

The approximate geometric ratios of the cavity/object were given by Reichardt (1946) and simplified by Garabedian (1956); its shape is approximately ellipsoidal:

$$\frac{L_C}{d} = \frac{C_D^{1/2}}{\sigma^{3/2}} \left[\frac{(\sigma + 0.008)}{(1 - 0.132\sigma^{1/7})^{1/2}(1.7\sigma + 0.066)} \right] \approx \frac{C_D^{1/2}}{\sigma} \left[\ln \left(\frac{1}{\sigma} \right) \right]^{1/2}$$

$$\frac{d_C}{d} = \frac{C_D^{1/2}}{(\sigma - 0.132\sigma^{8/7})^{1/2}} \approx \left(\frac{C_D}{\sigma} \right)^{1/2}.$$

Quite often the cavity length is assumed to be twice as large as the distance from the cavitator to the maximum cavity diameter. Note that other canonical axisymmetric shapes also follow power laws. This is shown pictorially in Figure 3 for a disk, for four cones, and for a sphere.

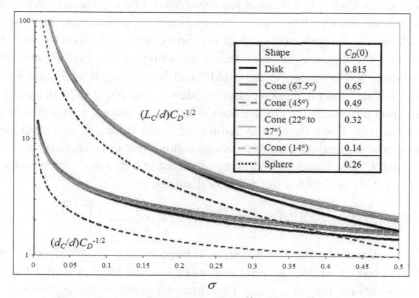

Figure 3: The average relationship between $(d_C/d)C_D^{-1/2}$ and $(L_C/d)C_D^{-1/2}$ for varying σ for the disk, cones with varying apex angles, and the sphere. The value of $C_D(0)$ is also given in the accompanying table reported in May (1975). The data for the sphere is derived Eqns. 7 and 8 using the experimentally measured value of $C_D(\sigma)$ averaged over varying Reynolds numbers (and hence different angles of cavity detachment). The value of $C_D(0)$ is also shown.

For a positive cavitation number, the pressure within the cavity is less than the pressure beyond, and hence the cavity closes in the direction of the object's longitudinal/flow direction axis. To describe this ellipsoidal-like shape, Münzer and Reichardt (1950) published it as

$$\left(\frac{2x - L_C}{L_C}\right)^2 + \left(\frac{2r}{d_C}\right)^{2.4} = 1.$$

Here x is the downstream distance, r is the cavity radius (that is a function of x), and the flow must be laterally unconfined (i.e. no walls present).

As might be expected, the most difficult part of the cavity problem is determining what occurs in the vicinity of the cavity closure. On the downstream side of a closed cavity, a stagnation point must exist. Although the cavity is closed, along its outer boundaries there must be downstream motion due to the drag from the outer flow, and hence there must be return flow within to enforce continuity. This type of "reentrant" jet closure is depicted in Figure 4. Reentrant jets are observed during natural cavitation as well as with ventilation and with super-cavities. Oftentimes, the closure itself does not significantly affect the cavity flow other than as regards the gas entrainment. In these cases, modeling of the cavities may be accomplished as shown in the middle and lower images of Figure 4 — the Riabouchinsky and open closure models — see Wu (1972). In cases of reentrant jets where the closure does affect the dynamics, very interesting and difficult flows develop including the cycle of filling and shedding partial cavities and pressure fluctuations within super-cavities. For example, Uhlman (2006) found the following relationship between the reentrant jet diameter, d_J, and that of the cavitator diameter, d:

$$\frac{d_J}{d} = \left(\frac{2C_D}{\pi}\right)^{1/2}\left[(1+\sigma)\left(1+\frac{1}{(1+\sigma)^{1/2}}\right)\right]^{-1/2}.$$

In cases where the flow speed, U, is large and the characteristic length of the object is small, the Froude number (i.e. the ratio of inertia forces to gravitational forces) is large. Therefore, the gravity effect becomes less important and hence so does buoyancy. Under these circumstances, one expects to see reentrant jets. In the real world, the reentrant condition

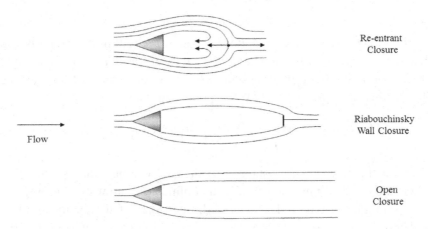

Figure 4: Schematic of the re-entrant flow at the closure of a cavity and a useful simplification used in many analyses, the Riabouchinsky wall closure model. Also shown is the open cavity closure for infinitely long supercavities (the open wake model).

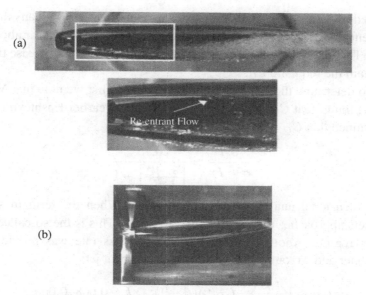

Figure 5: Examples of (a) reentrant closure with a close-up image of the jet (Schauer, 2003) and (b) twin vortex closure (Kunz *et al.*, 2001).

exhibits episodes of liquid filling the cavity followed by that liquid separating and flowing downstream. A reentrant jet is shown in Figure 5 with a magnified sub-region also shown. With moderate Froude numbers, on the other hand, there may be a profound influence of gravity on the cavity shape, especially with regards to its stabilization.

The gas in the cavity can be due to three factors: (1) vaporization (i.e. natural cavitation), (2) injection, and (3) dissolved gas diffused from the liquid into the cavity. The latter of these appears the least likely. In reality, very high speed flows and/or low ambient pressures cause natural cavitation while gas injection usually provides the ventilation route (although ventilation can occur directly from some flows such as beneath a planing craft).

To obtain natural cavities we offer some quantification. At 20°C, σ_V is 1.7 kPa (i.e. 0.25 psia). Hence assuming that the cavity length to disk cavitator ratio is $L_C/d > 20$, $\sigma < 0.05$ is required for cavitation. If one assumes that the ambient pressure is 100 kPa (i.e. atmospheric pressure), the flow speed required would be 60 ms^{-1}. On the other hand, assuming a speed of $10 \, \text{ms}^{-1}$ (such as one achieves in a laboratory) means that the ambient/freestream pressure needs to decrease below $10 \, \text{kPa}$ (or about 1.5 psia). This likely indicates that gas injection is required to increase the P_C to obtain the requisite L_C.

To determine the gas injection rate required, first we note that Michel (1984) found that $C_Q = C_Q \left[Fr, \beta \left(\equiv \frac{\sigma_V}{\sigma} \right) \right]$, and earlier Epshteyn (1961) determined that C_Q was

$$\frac{C_Q}{\sigma^{5/2} Fr} = \frac{0.75}{\left[\frac{\sigma Fr^{4/3}}{1.5^{2/3}} \right] - 1}$$

via dimensional analysis (and without σ_V!). When the term in square brackets approaches one, C_Q increases rapidly. This is the so-called twin vortex regime. Another estimate of the gas-loss rate was provided by Kirschner and Arzoumanian (2008) for a reentrant jet:

$$C_Q \approx k \frac{\pi}{4} \frac{1 + \sigma}{\sigma} C_D(0)(\beta - 1); \quad k \sim 0.008\text{--}0.009.$$

Further, known as the "stability" parameter, β is related to cavity oscillation, with $\beta = 1$ obviously for natural cavities and $\beta > 1$ for increased injected gas and the likelihood of decreased cavity oscillation.

Figure 6 (from Kirschner and Arzoumanian, 2008) with the upper inset from Schauer (2003) depicts downstream vortex shedding resembling a Karman Vortex Street, while the lower graph is from the authors' numerical simulations. At the trailing end of the cavity, the wave troughs are seen to pinch-off the cavity. In turn, the pinch-off causes pressure perturbations within the cavity that presumably couple into forcing the surface fluctuations. If the Froude number is such that gravity is important, these surface disturbances will cause three-dimensional structures.

As usual (cavity) disturbances may be damped or may grow and continue to oscillate. Silberman and Song (1961) found that if the stability parameter, β, was greater than 5.3 for 2D, cavities oscillate, while in 1978 Paryshev investigated stability theory via a dynamical systems approach that coupled the shape, volume, injection, entrainment rate, pressure, and drag. Linearizing the equations, he was able to determine that stable cavities occur for $P_O/(P_O - P_C) < 2.65$. Assuming that $P_O \gg P_V$ (i.e. that the ambient pressure is much larger than the vapor pressure), this ratio is

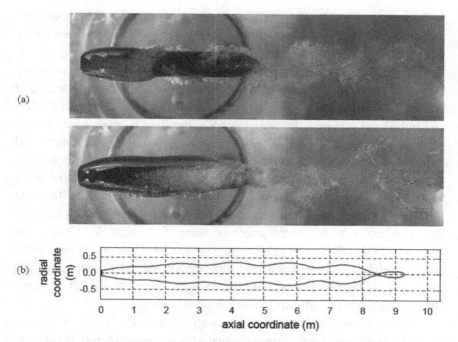

Figure 6: (a) Cavity vortex shedding and shape oscillation with visible re-entrant flow (Schauer 2003); (b) the computed 5th order mode of cavity oscillation for a cavity flow with $\beta = 93.9$. The waves intersect at the rear of the cavity, leading to the periodic pinch-off of gas bubbles (Kirschner & Arzoumanian 2008).

approximately β, while in a completely separate investigation, Kirschner and Arzoumanian (2008) found similarly for $\beta < 2.70$.

The solutions for the canonical cavity flows presented above are based largely on free-streamline theory, which captures a great deal of the cavity physics and is therefore quite useful when used as part of an overall stability analysis of the cavitating flow. Analysis of more complex and three-dimensional cavity shapes requires the use of numerical solutions. Brennen (1969) was one of the first researchers to employ numerical methods for the computation of the cavity. Uhlman (1987; 1989) discusses the use of Boundary Element Methods (BEM) for the computation of partially cavitating and super-cavitating two-dimensional hydrofoil flows. A similar approach was applied to axisymmetric super-cavitating flows by Uhlman *et al.* (1998), and to axisymmetric partially cavitating flows by Varghese *et al.* (2005). Kirschner *et al.* (2001) also review the use of slender body theory and BEM. Cavity flow calculations using these methods have captured successfully the steady and dynamic behavior of ventilated cavity flows.

Researchers have been working toward modeling of turbulent cavitating flows (natural and ventilated) through solution of the Reynolds Averaged Navier Stokes (RANS) equations with a number different physical models and numerical schemes. In most methods, the flow is modeled as homogeneous fluid with a variable mixture density. A single set of mass and momentum equations are solved for the mixture with a variable mixture density or void fraction, including turbulent transport. In the most basic formulation, the mixture density is governed by an equation of state that forces an abrupt change in density between the liquid and vapor phases as the local pressure crosses the vapor pressure (Song and He, 1998; Song and Qin, 2001). Wang and Ostoja-Starzewski (2007) report a Large Eddy Simulation (LES) employing a barotropic density model, for natural partial cavities. This method has been used for natural cavity flows, but it cannot capture the dynamics of injected non-condensable gas.

Homogenous flow models have been further developed to include multiple phases and dynamic mass transport between the phases. A separate volume fraction transport equation is solved that includes source terms for the dynamic exchange of mass between the phases (possibly as a result of phase change) using a Transport Equation based Model (TEM). Senocak and Shyy (2002; 2004a; 2004b) and Wu *et al.* (2005) use a pressure-based algorithm to solve the RANS equations for each phase, and both steady and unsteady natural cavitation are simulated. Figure 7 presents the results of a calculation of natural partial cavitation over a hemispherical headform using this method (Senocak and Shyy, 2004a). The figure shows the time-averaged mixture density computed using three different mass-transport models between the liquid and the vapor. The iso-contours of vorticity reveal recirculation in the cavity closure region that is indicative of re-entrant flow.

Kunz *et al.* (2000) and Venkateswaran *et al.* (2002) developed preconditioned time-marching algorithm for the computation of both compressible and incompressible multiphase mixtures. The finite compressibility of the pure gas-phase and the gas–liquid mixtures can be quite significant in cavitating flows. The sound-speed of low-void-fraction mixtures is reduced

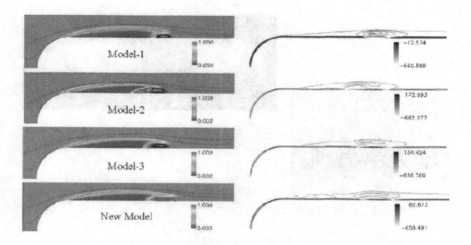

Figure 7: The results of a calculation of the time-averages natural, partial cavitation over a hemispherical headform (Senocak & Shyy 2004a). The density distribution (left) and the vorticity (right) computed using three different mass-transport models.

significantly from that of the pure liquid, and the motion of high-speed objects can approach sonic conditions of the bubbly mixture and, in some high speed cases, that of the pure liquid. The method developed by Venkateswaran *et al.* (2002) has been used to successfully compute a variety of canonical and practical, three-dimensional, unsteady, cavitating flows (Kunz *et al.*, 2000; 2001). Note that the method of Kunz *et al.* (2000) permits the simultaneous computation of the liquid, vapor, and non-condensable gas phases in the flow. Figure 8 presents sample results using this method as reported in Kunz *et al.* (2001), which show the gas and vapor flow near a cavitator with multiple injection ports, an unsteady natural cavity on an Ogive shape, and the flow around a notional underwater vehicle at an angle of attack.

The development and validation of these and other advanced computational methods is an ongoing topic of research, and researchers continue to explore new models for mass transfer and turbulent transport, both within multiple phases of the flow. This is particularly important to capture the gas exchange at cavity boundaries and the unsteady entrainment rate at turbulent cavity closures. Cavity flows may have sharp density gradients at the cavity interface that are well-captured with BEM or level-set methods. Conversely, the cavity closure is usually a turbulent, bubbly mixture,

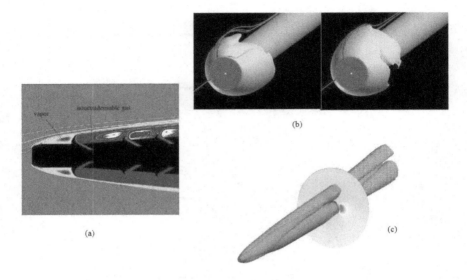

Figure 8: The results of calculation reported in Kunz et al. (2001) showing the gas and vapor flow near a cavitator with multiple injection ports (a); two realizations of an unsteady natural cavity on an ogive shape with the iso-surface of 90% void fraction highlighted for $\sigma = 0.30$ and Reynolds number of 1.46×10^{2} based on the diameter (b); the flow around a supercavitating object moving at a 5 degree attack angle with the 90% void fraction iso-surface and the magnitude of the velocity in a plane (c).

the physics of which are more readily captured with homogenous flow models. A combination of different modeling approaches may ultimately be necessary, as demonstrated by Chahine and Hsiao (2000).

At this juncture having covered the background materials for super-cavity flows, we revert to our primary focus in this chapter — to discuss gas injection for drag reduction for underwater vehicles. The overall effort has been the subject of investigations for decades, especially as it is known that the possibility of large increases in speed and/or in energy savings is conceivable through high drag reduction. Many groups have contributed especially in Russian and Ukraine; while in the United States, Pennsylvania State University, Naval Undersea Warfare Center, Alion Science and Technology, and University of Minnesota all have active research programs.

We present some long-standing issues. In Figure 9, a canonical self-propelled underwater vehicle is depicted with its salient features: (1) cavitator, (2) injectors, (3) control surfaces, and (4) planing surfaces. We assume that rocket propulsion is used. Alternatively, a propeller could be used, although then you would want the super-cavity to reduce to a partial-cavity so that the propeller would operate in liquid not gas. For speeds in excess of $70\,\mathrm{ms}^{-1}$, the cavity would likely envelop the entire body (i.e. a super-cavity), while at lower speeds a partial cavity with its closure approaching the stern and a cavitating wake would be more likely.

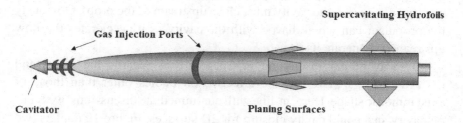

Figure 9: A schematic drawing of a canonical, self-propelled cavitating object. The basic hydrodynamic elements are the cavitator, a gas injector, planing surface, and control surfaces. Here, it is assumed that the object will be propelled by a rocket motor and be enveloped in a super-cavity. Alternatively, a rotating super-cavitating propulsor could be used for motive force. Then, the cavity would terminate upstream of the propulsor (e.g. a partial cavity) so that the blades of the propulsor would engage the liquid.

As might be expected, the design of the cavitator is fundamental to the success of the underwater vehicle. To insure a clean separation from the axisymmetric body, a circular disk is usually the geometric choice; additionally, it may be oriented so as to generate lift and/or lateral forces. Several other shapes have been used including (1) cones as in Figure 9, (2) disks with azimuthal variation, and (3) even cones with adjustable sizes (Ashley, 2001). As is shown also in Figure 9, and was mentioned previously, air injection is required, and may be necessary at multiple locations. The injection must be vectored in such a manner as not to overly perturb/destroy the cavity, and the gas can be compressed and stored on the object, or it can be generated chemically.

From a practical point of view, maneuvering presents a challenge. Planing surfaces known as "skids" may be located near the trailing end of the body to help avoid cavity breaching (i.e. to avoid extreme body motions). Also, lifting surfaces, or using the analogy with fish, fins may be used for control where these surfaces necessarily penetrate the cavity, but they must be designed for super-cavitation. Note that control of the vehicle is quite difficult as the center of pressure is on the disk cavitator itself.

As mentioned above, various methods of propulsion are possible. Propulsion is discussed in Savchenko (2001). Using the surrounding liquid (usually water) as the oxidizer and for example aluminum as the metal fuel, solid rockets may be used to generate thrust directly, or to make steam for a turbine. If the choice is made to drive a turbine rather than to use a direct jet, then a cavity breach propeller is used which necessarily requires a partial cavity (i.e. the cavity must close upstream of the prop). Obviously the propulsor can wreak havoc with the cavity as it accelerates the flow upstream to generate its thrust.

By way of our discussion in Chapter 5 as regards flat plates and the HIPLATE, PCDR may be effective on bodies other than those of axisymmetric shape. Here, in-line with our immediate discussions, we focus on cavity shape and cavity closure for 2D surfaces. Figure 10 depicts two types of partial cavities, the latter of which, a backward facing step that closes on a "beach", was the cavity type used in the experiments by Lay *et al.* (2010) and by Makiharju *et al.* (2010; 2013) as presented in Chapter 5. The upper schematic of Figure 10 shows a cavitator in the form of a wedge, with a reattaching cavity on the flat, downstream surface. We assume that gravity is as shown in the figure, or equivalently that the plate is oriented horizontally.

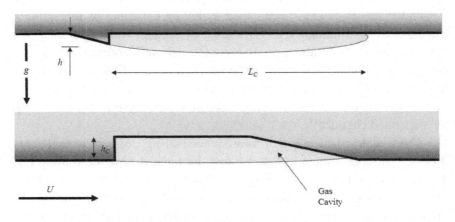

Figure 10: Two canonical partial cavity flows formed on two-dimensional surfaces. The cavity can separate from a protruding cavitator and then reattach to the surface (upper figure), or the cavity can form downstream of a backward facing step as seen in the lower figure.

Tulin (1957) presented power law expressions for the cavity shape on a 2D body much like those given for an axisymmetric body. The approximate expressions with h as the cavitator height and h_C as the cavity height, both as in Figure 10 are

$$\frac{L_C}{h} \approx \frac{16C_D(0)}{\pi\sigma^2}; \quad \frac{h_C}{h} = \frac{4C_D(0)}{\pi\sigma}.$$

Recall that $C_D(\sigma = 0)$ is the drag on a semi-infinite body, i.e. one that never closes. Note that the terms on the right-hand side are all dimensionless, so that the power of σ does not affect the dimensions. Tulin also derived an equation for small σ based on slender body theory for the ratio of the cavity length to height:

$$\frac{L_C}{h_C} = \frac{(2)(2+\sigma)}{\sigma}$$

which approaches the ratio above as $\sigma \to 0$. As for the axisymmetric vehicle, the cavities are ellipsoidal-like, and the closure model used, although important to the dynamics, does little to alter the cavity profile. In fact, using potential flow theory, Michel (1978) and Callenaere *et al.* (2001) solved for the cavity behind a backward facing step. The solution gives the entire cavity shape as well as its non-dimensionalized length and reentrant jet thickness:

$$\frac{y_C(r)}{h} = -(1-\lambda)\left[1 - \frac{1}{\pi}\cos^{-1}(2r-1) - \frac{2}{\pi}(r(r-1))^{1/2}\right]$$

$$\frac{x_C(r)}{h} = \frac{\lambda}{\pi}\left[\frac{1-2\lambda+2\lambda^2}{\lambda^2}r + \ln(1-r)\right]$$

$$\frac{L_C}{h} = \frac{\lambda}{\pi}\left[\frac{(1-\lambda)^2}{\lambda^2} + \ln\left(\frac{\lambda^2}{1-2\lambda+2\lambda^2}\right)\right]$$

$$\lambda = \frac{h_J}{h} = \frac{1}{2}\left[1 - \frac{1}{\sqrt{1+\sigma}}\right].$$

Here r is zero at the leading edge and one at the (infinite) extent of the reentrant jet. The latter two equations are presented in Figure 11 for the case of large Fr (i.e. $\gg 1$) while Figure 12 exhibits the cavity shapes from the two former equations for both large and small Fr.

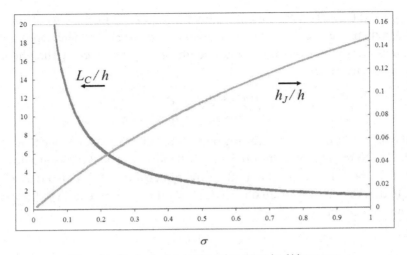

Figure 11: The length of the cavity and re-entrant jet thickness as a function of cavitation number for several values of σ for $Fr \gg 1$. In both of these cases, the re-attaching partial cavity will tend to form a re-entrant flow whose flux increases with increasing σ.

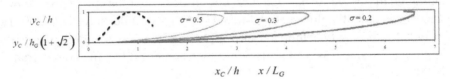

Figure 12: The cavity profiles, $y_C/h = f(x_C/h)$, for cavities forming at the base of a rearward facing step for $Fr \gg 1$ (solid lines) that show re-entrant flow, and for smaller Fr where the first wave of the cavity profile $y_c(x/L_G)/h_G(1+\sqrt{2}) = f(x/L_G)$ is shown (dashed line). The G subscript represents gravity-wave half-wavelength.

As mentioned for axisymmetric flows, cavity closure is very important especially for stability. And for two-dimensional cavities it has been investigated exhaustively. Several excellent reviews are available on this and related topics (e.g. Franc, 2001; Callenaere *et al.*, 2001; Kawanami *et al.*, 1998).

In cases where the Froude number is not large, gravitational effects are important and hence buoyancy will eventually dominate shape and closure. For cavities that also have $L_C/h \gg 1$, the cavity exhibits a gravity-wave type profile (as is the case for a transom stern at appropriate Fr — see Schmidt (1981) and Maki *et al.* (2008)). In Figure 12, one such profile is shown from the asymptotic theory of Schmidt (dashed line). Its equation is given in the figure caption. Lastly in the paper by Makiharju *et al.* (2013), other such profiles are presented.

To generate a long partial cavity (L_C/d or $L_C/h \gg 1$), injection is required. Closure is very important in terms of the gas flux required. There is a paucity of data on two-dimensional ventilated partial cavities. Laali and Michel (1984) investigated such a flow. The cavity length scaled similar to the equation introduced above, and the flow's entrainment rate ranged over $0.1 < C_Q < 0.7$ with the cavitation number, σ, within 0.01 and 0.15.

As shown in the previous chapter, and in Figure 10, a cavity can close on a sloped surface. And to reduce the possibility of reentrant flow as well as the injected gas requirement, a cavity should close with its surface oriented approximately aligned with the solid surface. Toward this end, Amromin *et al.* (2006), Kopriva *et al.* (2007; 2008) used a foil designed especially to generate a cavity that smoothly reattached near the midpoint of the body. Improvements in the lift-to-drag ratio of 30% to 50% were achieved for both natural and forced-injection cavities. They also investigated unsteady injection. For their two-dimensional cavities, $C_Q = O\,(0.005)$ with $C_Q \equiv Q/Uhw$ and w the span.

The experiments of Lay *et al.* (2010) and Makiharju *et al.* (2013) were discussed in Chapter 5 and the interested reader is referred there for details.

To conclude this chapter on super- and partial-cavities, we mention some complications of two-dimensional cavities. As the length of a sustainable cavity may be less than the surface over which drag reduction is desired, several cavities may be required. As is the case with cavities in general, the cavities are designed for a particular speed and may actually hinder operation at other speeds. In some instances, multiple downstream cavities may be designed such that as speed is increased, the cavities fill sequentially as depicted in Figure 13, thus facilitating multiple design speeds.

In addition, one must recognize the real-world effect of a cavity's span-wise extent. Transverse waves are possible, and as shown by Matveev (2007) as one might suspect, if $w > L_G$ (i.e. an aspect ratio of length to width < 1) especially, waves may occur. This increases the likelihood of producing reentrant jets. One way to avoid this is to shape the beach to inhibit reentrant flow (de Lange and de Bruin, 1998; Laberteaux and Ceccio, 2001).

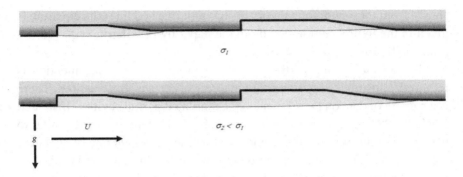

Figure 13: Schematic configuration of multiple step cavities. At higher σ (e.g. lower speed), each cavitator has a distinct cavity; at lower σ (e.g. higher speed), a single cavity grows to cover the two cavitators.

References

Amromin, E.L., Kopriva, J.E. and Arndt, R.E.A. (2006). Hydrofoil drag reduction by partial cavitation. *J. Fluids Eng.* 128(5), 931–936.

Ashley, S. (2001). Warp drive underwater. *Sci. Am.* 284, 70–79.

Brennen, C.E. (1969). The dynamic balance of dissolved air and heat in steady natural cavity flow. *J. Fluid Mech.* 37, 115–127.

Brennen, C.E. (1995). *Cavitation and Bubble Dynamics.* New York, Oxford Univ. Press

Callenaere M., Franc J.P., Michel J.M. and Riondet, M. (2001). The cavitation instability induced by the development of a reentrant jet. *J. Fluid. Mech.* 444, 223–256.

Ceccio, S.L. (2010). Friction Drag Reduction of External Flows with Bubble and Gas Injection. *Annu. Rev. Fluid Mech.* 42, 183–203.

Chahine, G.L. and Hsiao, C.-T. (2000). Modeling 3D Unsteady Seet Cavities Using a Coupled UnRANS-BEM Code. *23rd Sympos. on Naval Hydro.*

De Lange, D.F. and De Bruin, G.J. (1998). Sheet Cavitation and Cloud Cavitation, Re-Entrant Jet and Three-Dimensionality. *Appl. Sci. Res.* 58, 91–114.

Epshteyn, L.A. (1961). Determination of the amount of gas needed to maintain a cavity behind a body moving at low Fr number. *Tr. TsAGI* 824, 45–56 (In Russian).

Franc, J.-P. (2001). Partial Cavity Instabilities and Re-Entrant Jet. In: CAV 2001: *Fourth Int. Sympos. Cavitation*, California Institute of Technology.

Garabedian, P.R. (1956). The mathematical theory of three-dimensional cavities and jets. *Bull. Am. Math. Soc.* 62, 219–235.

Kawanami, Y., Kato, H. and Yamaguchi, H. (1998). Three-dimensional characteristics of the cavities formed on a two-dimensional hydrofoil. *Proc. Third Int. Sympos. Cavitation*, 191–196.

Kirchhoff, G. (1869). Zur Theorie freier Flüssigkeitsstrahlen. *J. für die reine und angewandte Mathematik* 70, 289–298.

Kirschner, I.N. and Arzoumanian, S.H. (2008). Implication and extension of Paryshev's model of cavity dynamics. *Proc. Int. Conf. Innov. Approaches Further Increase Speed Fast Mar. Veh., Mov. Above, Under Water Surf., FAST' 2008*, 1–32

Kirschner, I.N., Fine, N.E., Uhlman, J.S. and Kring, D.C. (2001). Numerical modeling of supercavitating flows. *Proc. RTO AVT Lect. Ser. Supercavitating Flows*, 9.1–9.39. RTO Lect. Ser. 005, RTO-EN-010.

Knapp, R.T., Daily, J.W. and Hammitt, F.G. (1970). *Cavitation*. New York, McGraw-Hill.

Kopriva, J.E., Amromin, E.L., Arndt, R.E.A, Wosnik, M. and Kovinskaya, S. (2007). High-performance partially cavitating hydrofoils. *J. Ship Res.* 51(4), 313–325.

Kopriva, J., Arndt, R.E. and Amromin, E.L. (2008). Improvement of hydrofoil performance by partial ventilated cavitation in steady flow and periodic gusts. *J. Fluids Eng.* 130(3), 0131301-1–0131301-7.

Kunz, R.F., Boger, D.A., Stinebring, D.R., Chyczewski, T.S. and Lindau, J.W. (2000). A preconditioned Navier-Stokes method for two-phase flows with applications to cavitation prediction. *Comput. Fluids* 29, 849–875.

Kunz, R.F., Lindau, J.W., Billet, M.L. and Stinebring, D.R. (2001). Supercavitation 3-D hydrofoils and propellers: prediction of performance and design. *Proc. RTO AVT Lect. Ser. Supercavitating Flows*, 13.1–13.44. RTO Lect. Ser. 005, RTO-EN-010.

Laali, A.R. and Michel, J.M. (1984). Air entrainment in ventilated cavities: Case of the fully developed "half-cavity". *J. Fluids Eng.* 106(3), 327–335.

Laberteaux, K.R. and Ceccio, S.L. (2001). Partial cavity flows. Part 1. Cavities forming on models without spanwise variation. *J. Fluid Mech.* 431, 1–41.

Lay, K.A., Yakushiji, R., Makiharju, S., Perlin, M. and Ceccio, S.L. (2010). Partial cavity drag reduction at high Reynolds numbers. *J. Ship Res.* 54(2), 109–119.

Maki, K.J., Troesch, A.W. and Beck, R.F. (2008). Experiments of two-dimensional transom stern flow. *J. Ship Res.* 52(4), 291–300.

Makiharju, S., Elbing, B.R., Wiggins, A., Dowling, D.R., Perlin, M. and Ceccio, S.L. (2010). Ventilated Partial Cavity Flows at High Reynolds Numbers. *Int. Conf. on Multi-phase Flows*, Tampa, FL.

Makiharju, S., Elbing, B.R., Wiggins, A.D., Schinasi, S., Vanden-Broeck, J.-M., Perlin, M., Dowling, D.R. and Ceccio, S.L. (2013). On the scaling of air entrainment from a ventilated partial cavity. *J. Fluid Mech.* 732, 47–76.

Matveev, K.I. (2007). Three-dimensional wave patterns in long air cavities on a horizontal plane. *Ocean Eng.* 34, 1882–1891.

May, A. (1975). Water entry and the cavity-running behavior of missiles. *SEAHAC Tech. Rep. 75-2*, Nav. Surf. Weapons Cent., White Oak Lab., Silver Spring, MD.

Michel, J.M. (1978). Demi-cavité formée entre une paroi solide et un jet plan de liquide quasi parallèle: approche théorique. DRME Contract 77/352, Rpt 4.

Michel, J.M. (1984). Some features of water flows with ventilated cavities. *J. Fluids Eng.* 106(3), 319–326.

Münzer, H. and Reichardt, H. (1950). Rotational symmetric source-sink bodies with predominantly constant pressure distributions. Aerospace Research Establishment, England.

Paryshev, E.V. (1978). A system of nonlinear differential equations with a time delay describing the dynamics of nonstationary axially symmetric cavities. *Tr. TsAGI* 1907 (In Russian).

Reichardt, H. (1946). The laws of cavitation bubbles at axially symmetric bodies in a flow. *Rep. Trans. 766, Minist. Aircr. Prod., Britain.*

Savchenko, Y.N. (2001). Supercavitating object propulsion. *Proc. RTO AVT Lect. Ser. Supercavitating Flows*, 17.1–17.29. RTO Lect. Ser. 005, RTO-EN-010.

Schauer, T.J. (2003). *An experimental study of a ventilated supercavitating vehicle.* MS Thesis. Univ. Minnesota.

Schmidt, G.H. (1981). Linearized stern flow of a two-dimensional shallow-draft ship. *J. Ship Res.* 25(4), 236–242.

Senocak, I. and Shyy, W. (2002). A pressure-based method for turbulent cavitating flow computations. *J. Comput. Phys.* 176, 363–383.

Senocak, I. and Shyy, W. (2004a). Interfacial dynamics-based modelling of turbulent cavitating flows, part 1: model development and steady-state computations. *Int. J. Numer. Methods Fluids* 44(9), 975–995.

Senocak, I. and Shyy, W. (2004b). Interfacial dynamics-based modelling of turbulent cavitating flows, part 2: time-dependent computations. *Int. J. Numer. Methods Fluids* 44(9), 997–1016.

Silberman, E. and Song, C.S. (1961). Instability of ventilated cavities. *J. Ship Res.* 5, 13–33.

Song, C.S. and He, J. (1998). Numerical simulation of cavitating flows by single-phase approach. *Proc. Third Int. Sympos. Cavitation*, 295–300.

Song, C.S. and Qin, Q. (2001). Numerical simulations of unsteady cavitating flows. *Proc. Fourth Int. Sympos. Cavitation*, 1–8.

Tulin, M.P. (1957). The theory of slender surfaces planing at high speeds, *Schiffstechnik Bd. 4, Heft* 21, 125–133.

Uhlman, J.S. (1987). The surface singularity method applied to partially cavitating hydrofoils. *J. Ship Res.* 31(2), 107–124.

Uhlman, J.S. (1989). The surface singularity or boundary element method applied to supercavitating hydrofoils. *J. Ship Res.* 33(1), 16–20.

Uhlman, J.S., Varghese, A.N. and Kirschner, I.N. (1998). Boundary-Element Modeling of Axisymmetric Supercavitating Bodies. *Proc. 1st HHTC CFD Conference*, Naval Surface Warfare Center Carderock Division, Carderock, MD.

Uhlman J.S. (2006). A note on the development of a nonlinear axisymmetric reentrant jet cavitation model. *J. Ship Res.* 50(3), 259–267

Varghese, A.N., Uhlman, J.S. and Kirschner, I.N. (2005). High-speed bodies in partially cavitating axisymmetric flow. *J. Fluids Eng.* 127, 41–54.

Venkateswaran, S., Lindau, J.W., Kunz, R.F. and Merkle, C.L. (2002). Computation of multiphase mixture flows with compressibility effects. *J. Comput. Phys.* 180, 54–77.

Wang, G. and Ostoja-Starzewski, M. (2007). Large eddy simulation of a sheet/cloud cavitation on a NACA0015 hydrofoil. *Appl. Math. Model.* 31, 417–447.

Wu, J., Wang, G. and Shyy, W. (2005). Time-dependent turbulent cavitating flow computations with interfacial transport and filter-based models. *Int. J. Numer. Methods Fluids* 49, 739–761.

Wu, T.Y. (1972). Cavity and wake flows. *Annu. Rev. Fluid Mech.* 4, 243–284.

Chapter 7

Super-Hydrophobic Surfaces and Coatings

Since Onda *et al.* (1996) demonstrated the properties of super-hydrophobic surfaces (SHS), there has been a considerable amount of research into the physics and applications of these surfaces. These surfaces have novel properties that have been exploited to repel liquids from them (including cloth surfaces, for example). Our discussion here will focus on the potential use of SHS to reduce drag developed by flowing liquids.

Definitions of super-hydrophobicity vary, but the fundamental idea is that the apparent contact angle, θ_c, between the liquid and the substrate exceeds some large value, usually $\theta_c > 150°$. Other important factors regarding SHS as it relates to drag reduction include contact angle hysteresis (the angular difference between the receding, θ_R and the advancing, θ_A contact angles), and low sliding angles. Note that hydrophilic surfaces are defined usually as those in which the contact angle $\theta_c < 90°$ and hydrophobic surfaces those with $\theta_c > 90°$.

To define angles more precisely, we present some simple sketches (For a more thorough discussion of contact angles, see e.g. the text by A.W. Adamson, *Physical Chemistry of Surfaces*, 1990.). When a drop of liquid is placed on a solid, the contact angle is defined as shown in the figure below. A liquid "wets" a solid if $\theta_c \to 0$; most liquids do not wet solid surfaces.

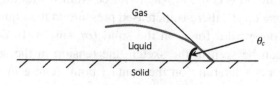

Two important equations come into play for contact angle problems: the Young–Laplace equation and Young's equation (the latter equation was published also by Laplace). In the Young–Laplace equation, $\Delta p = \sigma_{lg}\left(\frac{1}{R_1} + \frac{1}{R_2}\right)$, σ_{lg} is the surface tension and R_1 and R_2 are the radii of curvature of the surface. The equation gives the pressure jump, Δp across a fluid–fluid interface as a function of the surface curvature. A simple way to recognize the side with the higher pressure is that the pressure required to displace the surface, i.e. on the concave side, must be larger (Hence e.g. beneath a capillary wave crest, the pressure is larger.). Also, the subscript lg denotes the jump across the liquid–gas interface.

Young's equation, $\sigma_{lg} \cos \theta_c = \sigma_{sg} - \sigma_{sl}$, results from a balance of forces at the point of contact of the liquid, gas, and solid. We obtain this from the figure below where we assume a static drop, and that the forces can be represented by surface tensions (Young's equation can be derived also from energy considerations, thus substantiating the second statement above.).

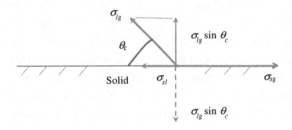

The along-solid forces are $\sigma_{sg} = \sigma_{sl} + \sigma_{lg} \cos \theta_c$ (i.e. Young's equation) while that perpendicular to the solid surface is simply $\sigma_{lg} \sin \theta_c$. Therefore to have equilibrium, requires a vertical downward force on the solid equal to this, and this is indicated by the dashed line in the figure. An alternative way to think of this is that according to the Young–Laplace equation, as the surface of the liquid is concave when viewed from within the liquid (as in a capillary wave crest), there is increased pressure in the liquid that causes an increased downward force on the solid ($\sigma_{lg} \sin \theta_c$). In the horizontal direction as can be seen in the sketch, the tension in the surface of the liquid clearly pulls laterally on the contact point (curve in 3D) and that must be balanced so that it remains pinned there; thus there must be a σ_{sg} oppositely directed with magnitude $\sigma_{sl} + \sigma_{lg} \cos \theta_c$.

To complete the definitions, the advancing (θ_A) and the receding (θ_R) contact angles are defined from work by Ting and Perlin (1995) where the authors investigated contact lines and contact angles on oscillating plates, and based their definitions on the uni-directional relative velocities (V_r) of Dussan (1979). θ_R, the receding angle is the contact angle as the contact point or line moves toward the liquid as $V_r \rightarrow 0$. Here V_r is defined as the velocity of the liquid relative to the solid. θ_A, the advancing angle is the contact angle as the contact point or line moves away from the liquid as $V_r \rightarrow 0$. Note that for all $V_r = 0$, the contact point is pinned to the solid, but θ_c is changing; this is the region of hysteresis. Shown in the Figure 1.3 is the contact angle versus relative velocity for uni-directional flows.

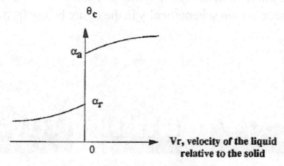

Figure 1.3 Contact angle versus relative velocity for uni-directional flow, showing hysteresis when $V_r = 0$ (From Dussan V. 1979).

Returning to SHS, we note that there have been several reviews of the subject, including that of Ma and Hill (2006) who investigated the methods of producing these surfaces (including those purpose-made to attempt self-cleaning such as for solar panels, and those produced to study how morphology affects contact angles and sliding angles); Roach *et al.* (2008) who also reviewed several techniques of manufacturing the surfaces, especially the progress toward artificial lotus-leaf-type surfaces; and Rothstein (2010) who discusses the use of roughness and contact angle to create surfaces that exhibit large slip. In general, these surfaces have many uses including: (1) self-cleaning as well as reduced drag in microfluids;

(2) various piping systems; (3) marine vessels; and even (4) anti-fogging coatings on eye-glasses.

While hydrophobicity can be achieved with smooth surfaces due solely to the surface energy at the interface of the solid, liquid, and gas, super hydrophobicity is often the result of both surface energy and surface morphology. For example, SHS can be realized if voids are captured on rough or textured surfaces. We thus define the Wenzel and the Cassie–Baxter states. These states are fundamental to the super-hydrophobic notion, were advanced more than 70 years ago, and were published prior to Onda *et al.* (1996). The two idealized topologies of a drop on a textured surface (Wenzel and Cassie–Baxter) simply stated are: (1) liquid fills the roughness on the surface, the Wenzel (1936) state, reducing θ_C; and (2) gas fills the crevices on the surface, the Cassie–Baxter (1944) state, and large contact angles result. There is of course a combined state also. All three of these wetting states are shown schematically in the figure below from Roach *et al.* (2008).

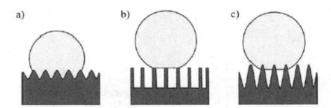

Fig. 1 Wetting states (a) Wenzel, (b) Cassie–Baxter and (c) combined models.

The authors (Roach *et al.*, 2008) proceed to discuss methods available for surface generation of SHS, many of which mimic nature. They discuss (1) fibers and textiles; (2) phase separation of mixtures; (3) crystal growth (some of these are even transparent); (4) differential etching; (5) diffusion-limited growth processes; (6) lithography (i.e. transferring patterns/designs to a surface by (a) contact (micron and nanometer sizes), (b) photolithography — irradiating a photoactive layer covered by a patterned mask — with UV, X-ray, direct etching by laser); (7) aggregation of particles, for example, by spin coating; (8) templating, that is using a template pattern and growing against it; and (9) using the aforementioned techniques to generate multi-scale roughness.

In Ma and Hill (2006), there is discussion as to whether the two roughness scales on SHS, that are micron scale (i.e. the rough scale) and nanometer scale (i.e. the small scale) can lead to highly effective and robust surfaces. In fact they cite natural examples that are SHS as well as two studies where nanometer-made surfaces exhibited the features of SHS, namely high contact angles and low sliding angles, such as the lotus leaf. Although they state that very low surface energies are not a requirement for non-wetting, they divide their paper into two primary techniques for the manufacture of SHS's: (1) using a low surface energy material and making it rough and (2) vice versa — creating a rough surface and then using low surface energy[1] material to alter it. Polymers (fluorinated ones in particular) and PolyDiMethylSiloxanes (PDMS) are low surface energy materials often used. Films of these substances are applied to a substrate. Roach *et al.* (2008) also focus their review on how the various surfaces can be manufactured.

Returning now to drag reduction, we ask the question: Can SHS's reduce friction drag of flowing liquids? For laminar flows, SHS is one of the only methods readily available to reduce drag. In the presence of a Cassie–Baxter state, the gas occupying the surface indentations is believed to lead to a reduced shear on the surface compared to a hydraulically smooth surface. This reduction is assumed to occur due to partial liquid slip along the surface of the solid (i.e. where the gas exists along the surface asperities).

One question that remains and is largely unexplored is what happens to the Cassie–Baxter state under the action of a turbulent boundary layer (TBL). Can the Cassie–Baxter state exist when subjected to the pressure fluctuations and shear stresses in a TBL? Will the high speed, low-pressure flow remove the gas and cause a transition to the less desirable Wenzel state?

---

[1]The concept of surface energy is a thermodynamic one — and is related to surface tension. Following Adamson's text again, and using σ for the surface tension (i.e. force per unit length of the interface), the work done to stretch a surface, W, is $W = \sigma l dx$ where dx is the distance extended, l the distance in the other surface dimension. Recall that work and energy (E) are dimensionally equivalent, that $l dx$ is an area, and this equation can be rewritten as $E = \sigma d(lx) = \sigma d(area)$, and in this latter expression σ is energy per unit surface area. Surface tension in MKS units is Nm^{-1} while energy is Jm^{-2}; they are dimensionally equivalent. And, the total surface energy is $E^S = \sigma - T\frac{d\sigma}{dT}$ with T the temperature.

These questions remain largely unanswered, and so we now discuss the drag reduction successes in laminar flows; we follow Rothstein (2010) wherein the discussion is focused on using surface roughness and engineered surfaces to cause large contact angle and small hysteresis, and hence increase/decrease slip/drag.

Following Rothstein (2010), it is accepted that for solid–liquid inter-faces, no-slip is the appropriate boundary condition to use. For certain flows that contain singularities such as an advancing free surface along a solid, there is theoretically an infinite, obviously non-physical stress produced (Dussan, 1979); it can be removed by using the slip model of Navier for example as done by Hocking (1976). On the other hand, for gas moving over a solid surface, Maxwell showed that the slip length is of the order of the mean free path (see for example a discussion based on the kinetic theory in Panton (1996) that provides the slip length). For most continuum mechanics flows, the slip length may be ignored.

An increase in the slip length corresponds to a reduction in the wall shear stress. There have been a limited number of measurement of slip lengths recently using techniques such as microPIV for the flow of water in a hydrophobic microchannel (Joseph and Tabeling, 2005). In this work, they measured a slip length of approximately 30 nanometers.

In Ou *et al.* (2004) and Ou and Rothstein (2005), it was shown that drag reduction could be achieved for laminar flows over super-hydrophobic surfaces. Defining w as the air width in the Cassie–Baxter state, h as the depth of the recesses, and d as the width of the solid surface between air regions (i.e. a square-wave surface with a duty cycle $d/(d + w)$ and amplitude, $h/2$), these researchers used lithography to etch and to self-assemble silicon to generate these square-wave shaped microchannels. Their drag reduction results gleaned from Rothstein (2010) are shown here.

As can be seen in their Figure 4b, they achieved DR's to approximately 40%, this for a dimensionless shear-free area (their $1 - \phi_s$ shown on the abscissa, the amount of air-water interface for posts rather than channels of $1 - d^2/(d + w)^2$) of 0.8. Interestingly, as the air exposure to total surface area increases, so does the drag reduction — in some sense this is similar to ALDR — and thus should not be a surprise.

In Ou and Rothstein (2005), the velocity profile was measured by microPIV. Their channel experiment had a glass upper surface while the

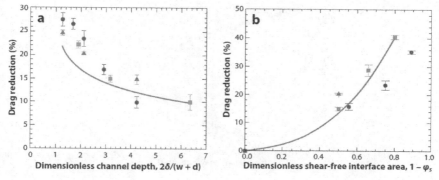

Figure 4

The average drag reduction as a function of (*a*) dimensionless channel depth and (*b*) percentage of shear-free interface. The data in panel *a* include superhydrophobic surfaces with a regular array of 30 μm wide microridges spaced 30 μm apart (*red triangles*), 20 μm wide microridges spaced 20 μm apart (*blue squares*), and 30 μm square microposts spaced 30 μm apart (*purple circles*) for a variable height microchannel. In panel *b*, the features are the same size, but the spacing between them is varied. Additionally, the microchannel is 2.54 mm wide, *H* = 127 mm high, and 50 mm long. Superimposed on both data sets are the predictions of computation fluid dynamics simulations for the 20 μm wide microridges spaced 20 μm apart case (*gray line*). Figure adapted with permission from Ou and Rothstein, 2005, Direct velocity measurements of the flow past drag-reducing ultrahydrophobic surfaces, *Phys. Fluids*, 17:103606, Copyright 2005, American Institute of Physics and reprinted with permission from Ou, Perot, and Rothstein, 2004, Laminar drag reduction in microchannels using ultrahydrophobic surfaces, *Phys. Fluids*, 16:4635–60, Copyright 2004, American Institute of Physics.

lower surface was composed of the microchannels. In their Figure 5 reproduced below from Rothstein (2010), one set of their measured velocity profiles is shown, and it is presented along with three velocity profiles: one to the top of the posts and two measured to the top of trapped air with different spacing. These data substantiated the drag reduction.

As reported in Rothstein (2010), Martell *et al.* (2009a; 2009b) used Direct Numerical Simulation (DNS) to numerically investigate *turbulent* flow over microchannel configurations. In these calculations, $y = 0$ was mid-channel, although it was defined differently elsewhere in the paper. One side of the flow, $y = -\delta$, was modeled as a SHS, the other as a no-slip surface. These are the solid curves shown in the Figure 5. Clearly good agreement is evident.

In Daniello *et al.* (2009), a PDMS super-hydrophobic surface and a no-slip surface were simulated. Their resulting velocity profiles on the SHS side and their friction coefficients are shown in the figure (their Figure 8).

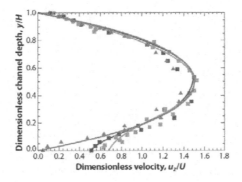

Figure 5

Velocity profiles measured through micro-particle image velocimetry (μ-PIV) for the flow through an $H = 85$ μm tall microchannel past a series of superhydrophobic surfaces containing $w = 30$ μm wide microridges spaced $d = 30$ μm (*red and purple symbols*) and $d = 60$ μm (*blue symbols*) apart. The data include the velocity profile for a vertical slice taken above the center of the microridge (*red triangles*), above the center of the 30 μm shear-free interface (*purple squares*), above the center of the 60 μm shear-free interface (*blue squares*), and the corresponding predictions of the computational fluid dynamics simulations (*lines*). Figure reprinted with permission from Ou & Rothstein 2005, Direct velocity measurements of the flow past drag-reducing ultrahydrophobic surfaces, *Phys. Fluids*, 17:103606, Copyright 2005, American Institute of Physics.

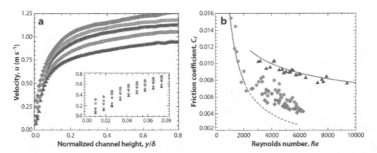

Figure 8

Velocity profiles and friction coefficient measurements for channel flow past superhydrophobic surfaces. (*a*) The Reynolds number increases from $Re = 5300$ (*dark blue triangles*) to $Re = 8000$ (*red diamonds*) over 30 μm wide ridges spaced 30 μm apart. (*b*) Reduced friction coefficients are noted for channel flow over superhydrophobic surfaces, with 30 μm wide ridges spaced 30 μm apart (*gold circles*) and 60 μm wide ridges spaced 60 μm apart (*red diamonds*). Smooth surface data (*dark blue triangles*) and predicted smooth values (*lines*) are shown for comparison. Figure adapted from Daniello R, Waterhouse NE, Rothstein JP. 2009. Turbulent drag reduction using superhydrophobic surfaces. In press in Phys. Fluids 2009 © American Institute of Physics, reproduced with permission.

   In the right inset the dark triangle symbols represent a no-slip smooth surface, while the lighter, filled circles and dark diamond symbols represent the 30 and 60 micron PDMS SHSs. The two curves shown are the laminar and turbulent C_f predictions, respectively. Evident in the figure is that transition occurs for the PDMS surfaces around a Re of 3,000 with Re defined as HU/ν.

In general, drag reduction beneath TBLs for these surfaces has not yet been definitively shown although Rothstein has given preliminary evidence of such results. This is a topic of considerable current interest, with researchers examining both the underlying physical processes and potentially effective material and morphological strategies. However, definitive measurements of reduced friction on test articles coated with SHS's are still unavailable. Yet, there are indications that SHS's will modify a TBL. For example, Aljallis *et al.* (2013), report on measurements of friction modification on a flat plate model covered in SHS nanoparticles and towed in water at relatively high Reynolds number of up to 10^7. They demonstrated FDR for both laminar and transitional boundary layers when the Cassie state was thought to be maintained. However, the FDR was lost when a fully TBL was developed over the plate, and this was ascribed to the loss of the trapped gas in the surface pores. In fact, the friction of the turbulent flow was somewhat higher than that of a baseline smooth flat plate, as the SHS increased the surface roughness.

Lastly, although we do not discuss in detail other passive surfaces that are not applicable to surface and underwater vessels, we mention here of another technique. As is known from the Olympic competitions in swimming, swim suit manufacturers have produced suits that apparently reduce drag. Therefore, we mention two papers related to this subject, one on swim suits themselves (Mollendorf *et al.*, 2004) and the other that reviews shark skin surfaces/riblets (Dean and Bhushan, 2010).

References

Adamson, A.W. (1990). *Physical Chemistry of Surfaces*. 5th edition, Wiley Interscience, New York.

Aljallis, E., Sarshar, M.A., Datla, R., Sikka, V., Jones, A. and Choi, C.H. (2013). Experimental study of skin friction drag reduction on superhydrophobic flat plates in high Reynolds number boundary layer flow. *Phys. Fluids* 25, 025103-1–025103-14.

Cassie, A.B.D. and Baxter, S. (1944). Wettability of porous surfaces. *Trans. Faraday Soc.* 40, 546–551.

Daniello, R.J., Waterhouse, N.E. and Rothstein, J.P. (2009). Drag reduction in turbulent flows over superhydrophobic surfaces. *Phys. Fluids* 21(8), 085103-1–085103-9.

Dean, B. and Bhushan, B. (2010). Shark-skin surfaces for fluid-drag reduction in turbulent flow: A review. *Phil. Trans. Roy. Soc. A* 368, 4775–4806.

Dussan, V.E.B. (1979). On the spreading of liquids on solid surfaces: Static and dynamic contact lines. *Annu. Rev. Fluid Mech.* 11, 371–400.

Hocking, L.M. (1976). A moving fluid interface on a rough surface. *J. Fluid Mech.* 76(4), 801–817.

Joseph, P. and Tabeling, P. (2005). Direct measurement of the apparent slip length. *Phys. Rev. E* 71, 035303.

Ma, M. and Hill, R.M. (2006). Superhydrophobic surfaces. *Current Opinion in Colloid & Interface Science* 11(4), 193–202.

Martell, M.B., Perot, J.B. and Rothstein, J.P. (2009a). Direct numerical simulations of turbulent flows over superhydrophobic surfaces. *J. Fluid Mech.* 620, 31–41.

Martell, M.B., Rothstein, J.P. and Perot, J.B. (2009b). The effect of Reynolds number on turbulent flows over superhydrophobic surfaces. *Phys. Fluids* 22, 065102-1–065102-13.

Mollendorf, J.C., Termin II, A.C., Oppenheim, E. and Pendergast, D.R. (2004). Effect of swim suit design on passive drag. *Med. Sci. Sports Exercise* 36, 1029–1035.

Onda, T., Shibuichi, S., Satoh, N. and Tsujii, K. (1996). Super-water-repellent fractal surfaces. *Langmuir* 12(9), 2125–2127.

Ou, J., Perot, J.B. and Rothstein, J.P. (2004). Laminar drag reduction in microchannels using ultrahydrophobic surfaces. *Phys. Fluids* 16, 4635–4660.

Ou, J. and Rothstein, J.P. (2005). Direct velocity measurements of the flow past drag-reducing ultrahydrophobic surfaces. *Phys. Fluids* 17, 103606-1–103606-10.

Panton, R.L. (1996). *Incompressible Flow*. 2nd edition, Wiley Interscience, New York.

Roach, P., Shirtcliffe, N.J. and Newton, M.I. (2008). Progress in superhydrophobic surface development. *Soft Matter* 4, 224–240.

Rothstein, J.P. (2010). Slip on superhydrophobic surfaces. *Annu. Rev. Fluid Mech.* 42, 89–109.

Ting, C.-L. and Perlin, M. (1995). Boundary conditions in the vicinity of the contact line at a vertically oscillating upright plate: An experimental investigation. *J. Fluid Mech.* 295, 263–300.

Wenzel, R.N. (1936). Resistance of solid surfaces to wetting by water. *Indust. Eng. Chem.* 28(8), 988–994.

Chapter 8

Passive Resistance-Mitigation by Appendages: Bulbous Bows, Stern Flaps and Wedges, and Lifting Bodies

In this chapter, we discuss appendages that serve to reduce the overall friction drag of an object moving in or on the surface of a liquid at high Reynolds number, such as ships and submarines. These appendages include bulbous bows, stern flaps, stern wedges, stern bulbs, lifting bodies, interceptors, and the like are application-driven and hence specific to a given hullform. Hence arriving at relevant canonical experiments are difficult. Yet, all these appendages are added to reduce propulsor demand or equivalently to increase speed, albeit in different ways.

Prior to proceeding with a discussion of these techniques, we cite relevant information from two papers by individuals associated with the U.S. Navy. The first of these is Doerry *et al.* (2010) that discusses energy costs and four ways that the U.S. Navy is struggling to contain/reduce these costs currently and in the future:

- Improved efficiencies of the prime mover.
- Decreasing the required propulsive power.
- Reducing required on-board power.
- Altering "modus operandi" as regards operations to reduce fuel consumption.

In this chapter, each is discussed in some detail; however, it is the second of these that is germane to this text.

To reduce the required power and hence save fuel (or equivalently increase speed for the same power input), the navy is implementing three primary (passive) techniques: (1) stern flaps, (2) bulbous bows, and (3) hull

135

and propeller coatings. Note that these methods are not new, although better implementations of them may now be available, and that these are primarily for navy displacement-hull vessels.

A stern flap is a nominally horizontal structural member that extends the bottom of the hull beyond the transom of the ship. Many variations of this method exist. By implementing the flap, the flow velocity is reduced locally, thus increasing the pressure in the vicinity, which increases the net forward force (as well as possibly reducing the trim), and thus reduces the form drag. A more in-depth discussion of the technical details and savings follows.

The discovery of bulbous bows is attributed to U.S. Admiral David Taylor at around the turn of the 20th century with the present-day implementation credited to Dr. Inui about mid-century. The purpose of the bulbous bow is to generate simultaneously out-of-phase waves with those of the ship bow-generated waves, thus helping to cancel each other by superposition, and ultimately reducing the energy input into the combined system. Obviously, less energy expended on wave generation means less power is required of the propulsion system for the same speed. With the advent of improved numerical models, bulbous bow designs have been improved such that they reduce power requirements over a larger range of ship speeds. In the case of military ships, a bulbous bow may be present for the sonar dome (to mask the ship's acoustic signature) rather than for energy reduction. In these cases, a second protuberance can be installed for the express purpose of reducing resistance. Again, an expansion of these ideas follows.

The third primary manner in which Doerry *et al.* (2010) addresses power reduction is the use of coatings on the hull and propeller. In Chapter 7, we discussed the use of super-hydrophobic surfaces while in Chapter 2, we discussed paints that release polymers, for example; Doerry *et al.* refer to coatings that are very smooth and resist fouling in various ways. Quite significant fuel savings can be achieved. The drawback of these techniques (as with all these techniques whether active or passive) is the increased cost and whether the cost–benefit ratio warrants pursuit of them. We do not discuss these further.

The second U.S. Navy publication/presentation we mention is that of Cusanelli and Karafiath (2012), and although their work and that of their collaborators is discussed in much more depth subsequently, their primary energy reduction efforts as regards hull hydrodynamics are similar to those mentioned above: retrofit bow bulbs (i.e. a second bulbous bow) to include a bulb on the stern and to improve the design of stern flaps themselves.

In the remainder of this chapter, we discuss bulbous bows first as they represent the oldest technology, followed by stern flaps and other stern appendages (i.e. wedges and bulbs). Lastly, we address the so-called lifting body methodology largely championed by Navatek, Ltd. These bodies generate sufficient lift to greatly reduce the hull wetted area thus reducing drag at least at higher speeds, and hopefully offering savings in spite of the increased drag on the lifting body.

8.1. The Bulbous Bow

As mentioned above, the effective design of a bulbous bow requires individual consideration of the specific hullform. This, in turn prevents us from developing a useful canonical geometry. Therefore, the results presented here are necessarily ship specific; however, they are indicative of the savings possible.

Following the text "Principles of Naval Architecture, Volume II, Resistance, Propulsion, and Vibration" (Lewis, 1988), also known simply as PNA, we review the basics of bulbous bows. Interested readers desiring additional information are referred to this text and its references as a starting point. It was David Taylor who first recognized that a wave generated by a submerged structure could cancel somewhat the wave generated by the bow of a ship, thus decreasing resistance (The so-called "ram" bow that was present on the HMS Leviathan, to ram ships, was the ship on which it was reported that Taylor first realized this phenomenon.). In Taylor's design, the ram bow was further submerged and broadened, and the notion of the bulbous bow was created. As far as an analytic theory was concerned, Havelock (1928) derived an expression for the surface waves about a sphere near the free surface of a constant freestream flow, followed by Wigley (1935–1936) who used Havelock's solution to publish "The Theory of the Bulbous Bow and its Practical Application" in which he calculated wave profiles and hence the related wave resistance.

For small Fr, Wigley actually saw increased resistance, however at higher Fr, resistance decreased. There are general guidelines available using the two coefficients introduced by Taylor, f and t, and they are discussed in PNA. Among Wigley's conclusions, the fact that drag reduction due to bulbous bows is present for $\sim 0.24 \leq Fr \leq 0.57$ is perhaps the most helpful. Bulbs have been shown to significantly reduce resistance, by as much as 20%, but usually more in the vicinity of 10%.

The next major step forward as far as this technique is concerned was made by Professor T. Inui in the early 1960's (Inui *et al.*, 1960). He used distributed singularities to determine an out-of-phase wave system, the summation of which with the bow-and-stern-generated waves produced a wave system containing much less energy. These basic ideas, although implemented now through numerical computations, are still in use today (For drag reduction, the U.S. Navy (see Cusanelli and Karafiath (2012), for example) is retrofitting ships with small bulbs located above the sonar/combatant domes. Again those interested in this technique are referred to the abundant literature on the subject.).

8.2. Stern Flaps, Wedges, and Bulbs

According to Karafiath and Fisher (1987), although trim flaps have been used on planing craft since the 1970's, stern wedges have seen no use on U.S. Navy displacement craft prior to 1987. There have been other navies that have used them sparingly. Interestingly, some German boats had used them prior to World War II while some Italian ships have implemented them more recently. Stern wedges are affixed to the bottom of existing transoms, while stern flaps extend beyond the transom, thus essentially extending the hull. In the paper by Karafiath and Fisher, the authors review historical data for seven configurations. They examine the effects on trim angle, wave resistance, effective power, and delivered power. More importantly, they present data for fuel savings and the required delivered power. Requisite delivered power reductions of ~ 6–7% were realized for a destroyer near its maximum speed. A new wedge presented exhibited a 2% savings in annual destroyer fuel costs.

Cusanelli and Hundley (1999) reported on stern flaps that were placed on a full-scale destroyer. Prior to this, the U.S. Navy had stern flaps installed only on a patrol boat and a frigate. These prototype alterations followed ~ 10 years of model tests. The basic design descriptors of stern flaps are the chord, span-wise extent, and the angle of the flap. As aforementioned, for sufficiently high speeds, they slow the flow aft, which in turn causes a pressure rise, thus reducing the net form drag. The flap may have adverse effects at slow speeds. According to Cusanelli and Hundley's full-scale results from the destroyer, the *A.W. Radford*, a delivered power reduction of 6 to 14% was achieved, or an increase in speed could be of ~ 1.5 ms^{-1}. These savings easily justified the cost to add the appendages. In the figure below, Figure 7 from their paper, the reduction of required power is presented as a function of ship speed.

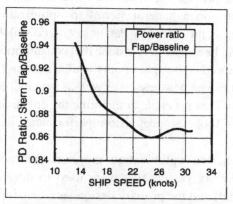

FIGURE 7. *A. W. Radford* reduction in delivered power, stern flap trial vs. baseline trial

Since those first two seminal publications, numerous publications have followed, particularly in the FAST (Fast Sea Transportation) Conferences held since 2003. We mention chronologically several of these contributions and their particular application vessels. Yaakob *et al.* (2004) investigated five configurations for reduced ship resistance on a planing hull crew boat (1:13.6 scale) model of the 34 m prototype. They found drag reduction for only one design, and its reduction was 7.2% for one speed. Fung *et al.* (2005) examined the vertical placement of stern flaps on a Series 64 hull as well as the effects of modifying its bow with two bulbous bow configurations and a so-called wave-piercing bow. Their results showed resistance savings of more than 10%. Exploring stern flap appendages for HALIFAX frigates, Cumming *et al.* (2006) found results in agreement with the literature: increased resistance at lower speeds, benefits at higher speeds. Cusanelli

(2011) published results for stern flaps on an amphibious vessel. As U.S. Navy amphibious ships include so-called well deck, which necessarily require folding stern gates to facilitate access, stern flap installations are unavoidably more complex. Cusanelli reported designs for three classes of ships as well as sea trials on a LHD 3, the *KEARSARGE*, with underway fuel savings of 3.4%. Day and Cooper (2011) studied "interceptors", which are slightly smaller than Gurney flaps, for drag reduction on high-speed sailing vessels. They stated that a 10 to 18% calm-water resistance decrease was achieved for speed of 8 to 20 knots.

As mentioned previously, canonical investigations are limited, and therefore we focus on a general stern-flap optimization technique for preliminary design as presented by Parsons *et al.* (2006). Owing to the fact that stern flaps have been shown to reduce fuel consumption or optionally to increase speed, the authors sought a path to initial designs (to be followed by model testing) for use on new vessels as well as for modification of existing craft. According to Parsons *et al.*, at that time the only existing preliminary stern flap design tool for propulsion performance was by Cusanelli (2002).

The multi-criterion optimization program incorporated the vessel parameters (of the six vessels that were used to develop the stern flap performance model) and the primary design variables for stern flaps into a model that would facilitate their use. As independent variables this program uses the volume *Fr*, the waterplane coefficient, the stern flap angle, and the chord length; all are defined in detail in the paper. The dependent variable was the ratio of the power delivered with the flap to the power delivered in the flap's absence (As the trim is more important for smaller ships, Parsons and co-workers added the waterplane coefficient.). Their final regression equation then was solved using the available data and based on levels of success, four regression models were included in their final product.

8.3. Lifting Bodies

A lifting body or hydrofoil can be used to remove a large portion of the hull from the water surface to reduce the wetted/liquid-contact area. These bodies have been used to improve seakeeping and drag simultaneously; however they may require active control.

The basic notion is that at slow speeds the vessel sits low in the water with added drag due to the appendages/foils. As the vessel's speed is increased, the lift generated by the appendages raises the vehicle significantly, even to a height such that only the appendages remain in the water. PNA terms this latter state as "foilborne" (analogous with airborne). And, as basic background, from that same text, we present their Figure 104 that illustrates the various stages of motion as well as an order of magnitude comparison of the contribution to the total resistance (ordinate axis does not include a scale) as a function of speed of the vessel.

In this figure, R_T is the total resistance, R_H is the hull resistance, D_{FL} is the foils' drag, D_{Str} is the struts' drag, R_{AP} is the appendages' drag, R_{AA}

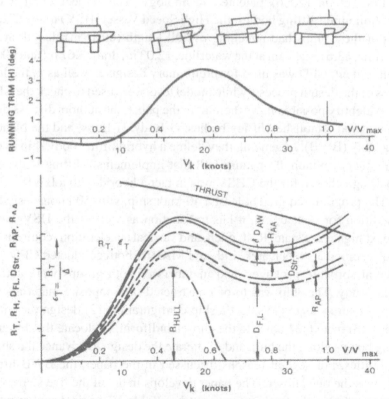

Fig. 104 Distribution of Resistance Components, Running Trim, Attitude of Hydrofoil Craft

is the wind resistance, and δD_{AW} is the increase in wave resistance. As the usual resistance is proportional to V^2, and that curve shape is seen for low speed, the reduction in resistance realized is obvious.

The savings however have associated costs in terms of the mechanical complexity of the hull, and control of the vessel becomes an important issue. We do not proceed to review the literature on hydrofoils or lifting bodies, but rather we concentrate on a relatively new series of vessels from the last decade.

In Hackett *et al.* (2007), the authors use CFD and model tests to investigate a high-speed vessel that used lifting bodies to reduce resistance and improve seakeeping. With funding from the Office of Naval Research under the Composite High Speed Vessel (CHSV) program, they explored the Loui *et al.* (2006) patented technology. The project compared a monohull minus lifting bodies, the High Speed Vessel (HSV) to the CHSV. Both of these ships had matching overall length (89.15 m); length at the waterline, 83.61 m; beam at the waterline, 12.03 m; draft, 3.61 m; and 2000 t displacement. CFD was used for preliminary design as well as in the later stages of the design process, while model tests were used to check the CFD, and to identify possible flaws therein. In the paper, the authors discuss three patented and established lifting bodies: G-Body, H-Body, and the blended wing body (BWB). In general, they define a hybrid lifting body ship as one with either a monohull or multi-hull that implements a lifting body. The ship design chosen for the CHSV was in fact a hybrid with a BWB.

The program required at least a 40-knot ship, with 50 knots desirable. In addition for seakeeping, it has to function as well as the HSV, but in the next higher sea state. Roll, pitch, and vertical acceleration requirements were specified also. Navatek, Ltd. and MARIN both conducted CFD with different software packages, and members of both groups were authors on the study. The ship was to be constructed of composite materials and the procedure was to (1) select a ship configuration, (2) design the lifting bodies, (3) mate the bodies to the chosen hullform, (4) locate the designed lifting bodies along the hull, and (5) iterate the design to advance the ship's capabilities. For several reasons discussed in the paper, the BWB lifting body was the one chosen. The paper develops in detail the five steps, with the figures below the final dimensions (their Table 4), the final configuration (their Figure 13), and a photograph of the model as tested (their Figure 14).

Table 4: CHSV Final Principal Characteristics Inclusive of BWBs

Parameter	Value
Length Overall (LOA)	90.60 m
Length on Waterline (LWL)	74.18 m
Beam Overall (BOA)	20.87 m
Beam on Waterline (BWL)	19.12 m
Draft, Design	3.68 m
Displacement, Design	1,995 t

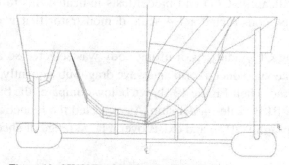

Figure 13: CHSV Final Design Configuration

Figure 14: CHSV Model

Comprehensive model tests were conducted and compared to the CFD results. Maneuvering as well as seakeeping and resistance tests were undertaken.

Of course in this text, we are interested primarily in resistance savings, and so we would like to concentrate on these results. Unfortunately, Hackett *et al.* focused on comparison of CFD to model tests (that were favorable including the CHSV resistance results), but a comparison of the CHSV resistance as to the HSV resistance was not provided.

In FAST 2007, Peltzer *et al.* (2007) published a paper on the bow lifting body (BLB) ship. Knowing that prototype ships using lifting bodies had demonstrated successfully improved seakeeping and reduced drag, they decided that perhaps including a lifting body at the bow (BLB, or bow lifting body) might serve two purposes: afford wave-canceling benefits as does a bulbous bow, and raise the ship relative to the water level thus providing the additional advantages of a usual lifting body. As each function serves to reduce a drag component, it was hoped that the total resistance could be reduced significantly. CFD and model tests indicated this to be the case; hence Navatek constructed a 1:4 scale demonstrator that was 21.3 m in length.

Following CFD analyses, a lifting body was selected so as to reduce total resistance via reduced form and wave drag. Subsequently, model tests were performed. Their Figure 13 shown below compares the bare hull drag to that of the BLB scale-model ship. As expected the reduction in drag is impressive (roughly 20%) and exists over a large range of speeds.

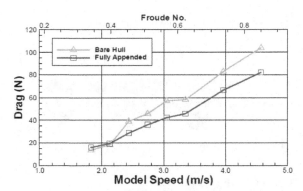

Figure 13 Reduction in Hump Speed Drag Validated

As mentioned, Navatek built a $^1/_4$-scale demonstrator called the BLB-70. It was put through comprehensive sea trials as they wanted to compare both to model and CFD calculations. Unfortunately the paper does not present the barehull results for the $^1/_4$-scale ship. Table 1 and Figure 15 from that publication reproduced here give the principal dimensions and a breakdown of the vertical loads carried by the two foils and the hull, respectively.

Table 1 BLB-70 Principal Characteristics

Parameter	Value	
Length Overall	69.90 ft	21.30 m
Beam Overall	19.00 ft	5.79 m
Length$_{WL}$, Main Hull	67.80 ft	20.67 m
Beam$_{WL}$, Main Hull	10.42 ft	3.18 m
Draft, Maximum	5.42 ft	1.65 m
Displacement, Design	22.0 LT	22.4 t
Installed Power	2 x 704 hp	2 x 525 kW
Speed, Design	23.6 knots	
Speed, Maximum	30+ knots	

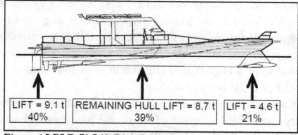

LIFT = 9.1 t 40%   REMAINING HULL LIFT = 8.7 t 39%   LIFT = 4.6 t 21%

Figure 15 BLB-70 Lift Distribution

In the FAST 2011 Conference, Engle *et al.* (2011) presented CFD seakeeping results and loads calculation for the full-scale BLB-160 ship. Once again, unfortunately resistance estimates were not presented.

We conclude this section (and chapter) on Lifting Bodies by mentioning that Hart *et al.* (2007) have conducted a seakeeping analysis of the "Sea Flyer", another hybrid lifting body ship converted from a surface effect ship by Navatek for the Office of Naval Research. Again, there are no resistance measurements or predictions presented of the barehull versus the fully-outfitted vessel.

References

Cumming, D., Pallard, R., Thornhill, E., Hally, D. and Dervin, M. (2006). Hydrodynamic design of a stern flap appendage for the Halifax Class Frigates. *Mari-Tech*, Halifax, N.S. Canada, June.

Cusanelli, D.S. (2002). Guidance for Implementing Stern Flap Performance Enhancements into ASSET Hullform Development Process. *NSWC Hydromechanics Report NSWCCD-50-TR-2002/052*, Aug.

Cusanelli, D.S. (2011). Hydrodynamic and Supporting Structure for Gated Ship Sterns — Amphibious Ship Stern Flap. *FAST 2011*, Hawaii USA, Sept.

Cusanelli, D.S. and Hundley, L. (1999). Stern flap powering performance on a spruance class destroyer: Ship trials and model experiments. *Naval Eng. J.* 111(2), 69–81.

Cusanelli, D.S. and Karafiath, G. (2012). Hydrodynamic Energy Savings Enhancements for DDG 51 Class Ships. *ASNE Day*, Arlington VA, USA, Feb.

Day, A.H. and Cooper, C. (2011). An experimental study of interceptors for drag reduction on high-performance sailing yachts. *Ocean Eng.* 38, 983–994.

Doerry, N.H., McCoy, T.J. and Martin, T.W. (2010). Energy and the Affordable Future Fleet. *10th Int. Naval Eng. Conf. Exhib.* (INEC), Portsmouth U.K., May.

Engle, A., Lien, V. and Hart, C. (2011). Seakeeping Evaluation and Loads Determination of a High-Speed Hull Form with and without a Bow Lifting Body. *FAST 2011*, Hawaii USA, Sept.

Fung, S.C., Karafiath, G. and Toby, A.S. (2005). The Effects of Bulbous Bows, Stern Flaps and a Wave-Piercing Bow on the Resistance of a Series 64 Hull Form. *FAST 2005*, St. Petersburg, Russia.

Hackett, J.P., St. Pierre, J.C., Bigler, C., Peltzer, T.J., Quadvlieg, F. and van Walree, F. (2007). Computational Predictions versus Model Testing for a High Speed Vessel with Lifting Bodies. *SNAME Annu. Meeting*, Nov. 1–18.

Hart, C.J., Weems, K.M. and Peltzer, T.J. (2007). Seakeeping Analysis of the Lifting Body Technology Demonstrator *Sea Flyer* Using Advanced Time-Domain Hydrodynamics. *FAST 2007*, Shanghai, China, Sept.

Havelock, T.H. (1928). The wave pattern of a doublet in a stream. *Proc. Roy. Soc. A* 121, 515–523.

Inui, T., Takahei, T. and Kumano, M. (1960). Wave profile measurements on the wave-making characteristics of the bulbous bow. *Trans. Japan Soc. Naval Arch.* 108.

Karafiath, G. and Fisher, S.C. (1987). The effect of stern wedges on ship powering performance. *Naval Eng. J.* 99(3), 27–38.

Lewis, E.V. (ed.) (1988). *Principles of Naval Architecture, Volume II, Resistance, Propulsion, and Vibration*. The Society of Naval Architects and Marine Engineers.

Loui, S., Shimozono, G. and Keipper, T. (2006). Low drag submerged asymmetric displacement lifting body. *US Patent 7,004,093 B2*.

Parsons, M.G., Singer, D.J. and Gaal, C.M. (2006). Multicriterion optimization of stern flap design. *Marine Tech.* 43(1), 42–54.

Peltzer, T.J., Keipper, T.S., Kays, B. and Shimozono, G. (2007). A New Paradigm for High-Speed Monohulls: The Bow Lifting Ship. *FAST 2007*, Shanghai China, Sept.

Wigley, C. (1935–1936). The theory of the bulbous bow and its practical application. *Trans. NECI* 52.

Yaakob, O., Shamsuddin, S. and King, K.K. (2004). Stern flap for resistance reduction of planing hull craft: A case study with a fast crew boat model. *J. Teknologi.* 41(A), 43–52.

Index

149

Printed in the United States
By Bookmasters